Praise for *Plant Science for Gardeners*

Robert Pavlis hit the nail on the head: know the science and you can grow anything. *Plant Science for Gardeners*, Pavlis' latest, is a sure (and enjoyable) way to learn that science. Mind you, Pavlis is not just a gardener who knows his science, he is a great science writer who does a fabulous job of making science fun. I am quite sure you are one read away from being a much better gardener.

—Jeff Lowenfels, author, *DIY Autoflowering Cannabis* and *Teaming with Microbes*

A work of genuine importance by an author for whom the phrase "received wisdom" is a contradiction in terms. You will never take horticultural lore at face value again.

—James Armitage, editor, *The Plant Review*, magazine of the Royal Horticultural Society

Knowing just how plants work is an interesting and useful way to ratchet up your gardening game and Robert Pavlis has provided that story in his engaging new book *Plant Science for Gardeners*. Read it and your garden will blossom, literally and figuratively.

—Lee Reich, author, *Growing Figs in Cold Climates* and *The Ever Curious Gardener*

Robert Pavlis has provided another detailed, yet accessible, addition to gardeners' home libraries. Be sure to give *Plant Science for Gardeners* an honored place on your bookshelves, next to Mr. Pavlis' groundbreaking *Garden Myths* series.

—Rebecca Martin, technical editor, *Mother Earth News* magazine

Plant Science for Gardeners is at once easy to read and comprehensive in presentation. Robert Pavlis has updated and made accessible information that in the past I have gleaned here and there from old textbooks and the odd gardening book. It's a great reference and a fascinating read. This book will be the next gift I buy for my budding botanist granddaughter!

—Darrell Frey, author, *The Bioshelter Market Garden*, co-author, *The Food Forest Handbook*

Whether you grow plants for fun or for profit, *Plant Science for Gardeners* does a great job of explaining how plants work without overcomplicating it. This is an excellent book not only for the basics, but also to explain some of the lesser known aspects of plants. Whether you've taken plant biology and physiology or not, it is a good read for anyone who's interested in plants, and to keep on the shelf as a reference for when you forget the difference between a node and internode. Get this book to understand what to do to keep your plants healthy and why.

—Andrew Mefferd, editor and publisher, *Growing for Market* magazine, author, *The Organic No-Till Farming Revolution*

plant science
for gardeners

plant science
for gardeners

Essentials for Growing Better Plants

ROBERT PAVLIS

Copyright © 2022 by Robert Pavlis. All rights reserved.

Cover design by Diane McIntosh.
Cover image © iStock

Printed in Canada. Second printing July 2024.

Inquiries regarding requests to reprint all or part of *Plant Science for Gardeners* should be addressed to New Society Publishers at the address below. To order directly from the publishers, please call 250-247-9737 or order online at www.newsociety.com.

Any other inquiries can be directed by mail to:
New Society Publishers
P.O. Box 189, Gabriola Island, BC V0R 1X0, Canada
(250) 247-9737

LIBRARY AND ARCHIVES CANADA CATALOGUING IN PUBLICATION

Title: Plant science for gardeners : essentials for growing better plants / Robert Pavlis.

Names: Pavlis, Robert, author.

Description: Includes index.

Identifiers: Canadiana (print) 20220154007 | Canadiana (ebook) 20220154031 | ISBN 9780865719736 (softcover) | ISBN 9781550927672 (PDF) | ISBN 9781771423632 (EPUB)

Subjects: LCSH: Botany. | LCSH: Gardening.

Classification: LCC QK50 .P38 2022 | DDC 580—dc23

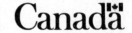

New Society Publishers' mission is to publish books that contribute in fundamental ways to building an ecologically sustainable and just society, and to do so with the least possible impact on the environment, in a manner that models this vision.

Contents

Introduction 1
  Introduction to Plant Science 2
  Organization of the Book 2
  Terms Used in This Book 3

1. Plant Basics 5
  Cells 5
  Xylem and Phloem 6
  Photosynthesis 7
  Plant Myth: Plants Raise the Oxygen Level in Homes . . . 8
  ATP and the Energy Cycle 10
  Respiration 11
  Meristematic Cells 12
  Classification of Plants 14

2. Roots 15
  Root Basics 16
  Types of Roots 19
  Plant Myth: Feeder Roots Are Located Under the Dripline . . . 20
  Effect of Gravity 21
  Cut Roots and Side Roots 22
  Plant Myth: Circulating Roots Continue to Circulate . . . 22
  Conditions That Affect Root Growth 23
  Plant Myth: Transplant Solutions Grow Better Roots . . . 24
  Root Growth in Winter 25
  Absorption of Water and Nutrients 26
  Plant Myth: Is Soil pH Important? 29
  Roots and Microbes 31
  Plant Myth: Purchased Mycorrhizal Fungi
    Are Good for Plants 32

Plant Myth: Roots Grow Towards Water	35
The Rhizosphere	35

3. Stems — 41

The Outer Structure of Stems	42
Internal Structure of Stems	42
Trichomes	44
Buds	44
How Stems Grow	45
The Importance of Photosynthesis	48
Underground Stems	49
Plant Myth: Plant Bulbs after the Ground Is Frozen	50
Plant Myth: Bearded Iris Should Be Planted with Rhizome Showing	52

4. Leaves — 55

Leaf Structure	55
How Sun Affects Leaves	60
Why Are Some Leaves Red?	62
Plant Myth: Evergreen Needles and Oak Leaves Are Acidic	63
Functionality of Damaged Leaves	63
Natural Pesticides	64
Signaling Between Plants	66
Pest-Proof Leaves	67
Water Stress and Wilting Leaves	67
Leaf Abscission	68
Variegated Leaves	71

5. Flowers — 75

Parts of a Flower	75
Pollination	77
What Causes Flowering?	78
Why Do Plants Not Flower?	80
Plant Myth: High Phosphate Grows More Blooms	81
Tough Love for Plants	84
Attracting Pollinators	85
Enjoy the Bracts	87
Dioecious and Monoecious Plants	88

6. Fruits and Seeds 91
   What Is a Fruit? 91
   The Importance of Fruit 92
   Different Types of Fruits 93
   Fruit Development 93
   Seed Development 95
   Suckering Tomato Plants 97
   Seeds from Non-Flowering Plants 98
   Soil Seed Bank 98

7. The Whole Plant 101
   Life Cycle of Plants 101
   Plant Dormancy 106
   Movement of Water 107
   Movement of Nutrients 109
   Plant Myth: Leaves Can Be Used to
      ID Nutrient Deficiencies 110
   Movement of Sugars 112
   Seasonal Sharing of Resources 113
   Overcoming Physical Damage 114
   How Do Plants Get Taller? 115
   Following the Sun 115
   How Light Affects Plant Growth 116
   Gravity 117

8. Woody Plants 119
   What Are Woody Plants? 119
   Structure of Woody Stems 120
   Where Does Wood Come From? 123
   Plant Myth: Newly Planted Trees Need to Be Staked . . . 124
   Storage of Sugars 125
   Taproots vs. Fibrous Roots 125
   Composition of Wood 127
   Apical Dominance 127
   Healing Damage 131
   Plant Myth: Damage on Trees Should Be Painted 132
   Conifers 133

Contents

9. Environmental Factors ... 137
    Garden Hardiness Zones ... 137
    Dealing with Cold ... 140
    Protecting Plants from Cold ... 142
    Dealing with Heat ... 145
    Dealing with Water Extremes ... 146
    Adaptability of Plants ... 149
    How Climate Change Affects Gardens ... 152

10. Growing from Seeds ... 157
    When Is Seed Mature? ... 157
    The Seed Germination Process ... 159
    The Mysterious Cotyledons ... 160
    Why Do Seeds Stay Dormant? ... 162
    Breaking Dormancy ... 164
    Plant Myth: Seeds Can Have Double Dormancy ... 164
    Seed Storage ... 167
    Best Method for Starting Seeds Indoors ... 168

11. Selecting Seeds ... 175
    Basic Genetics ... 175
    Hybrids vs. Heirlooms ... 176
    GMO Seeds ... 176
    Days to Maturity ... 177
    Buying Unusual Seeds ... 179

12. Vegetative Reproduction ... 181
    Natural Vegetative Reproduction ... 182
    Rooted Stems and Leaves ... 183
    Artificial Vegetative Reproduction ... 184
    Plant Myth: Homemade Rooting Hormones Work Well ... 191
    Grafting ... 192

13. Plant Names ... 195
    Why Use Botanical Names? ... 195
    Naming Conventions ... 196
    The Proper Way to Name Your Plants ... 198

Index ... 201
About the Author ... 209
Connect with Robert Pavlis ... 210
About New Society Publishers ... 212

*This book is dedicated to all the scientists
that toil long hours trying to elucidate the mysteries inside plants.
It is a slow and tedious process that requires dedication
and a true love of science. Without you, we would know
so little about the world around us.
It is my true hope that the general public
learns to appreciate scientists more.*

*I want to give a special thanks to Marika Li
who was instrumental in gathering information,
fact checking, and helping me prepare the manuscript.*

Introduction

New gardeners, and even more experienced ones, tend to learn about gardening by memorizing rules. When do you prune a lilac? Should you pinch out fall asters? When is the best time to move tulips? These all have rules, and once you learn them, they are easy to follow. Prune lilacs after flowering, pinch back asters in midsummer for stockier plants and transplant tulips once the leaves go yellow. But there are thousands of different kinds of plants. You will never learn and remember all the rules for all these plants.

A much better approach is to learn the underlying science—learn how plants grow and develop. Once you understand that, you can skip learning the rules because you don't need them and you will be able to grow just about anything. And that is the main goal of this book: I want you to understand what is really going on inside plants, and how they respond to the environment and your actions in the garden. This book paints a simple, clear picture of the greenery that surrounds you.

Once you have a really good understanding of the basics, you will be able to evaluate any gardening procedure and determine if it makes sense. For example, once you understand how dormant buds respond to pruning, you will have a much better understanding of when and what to cut. Learning about the transfer of water from roots to leaves will explain why some plants wilt at midday and what, if anything, you should do about it.

More importantly, you will be able to evaluate many of the fad techniques and products that are invented every year. Many of these are simply a waste of time and do not improve the health of your plants. This book will make you a more informed consumer.

Introduction to Plant Science

Plants have been studied for a few hundred years and we know a lot about them but with each advancement in our knowledge we realize there is so much more to learn. Plants are complex organisms and we are just starting to appreciate how they really work.

In order to write this book I had to make a lot of choices about what to include and what to leave out. If I'd included everything, this book would be ten times as big, and few people would read it. I have included basic information, like what is a leaf and how does pollination work, to give you a strong foundation of plant science. I then added other topics that are not only interesting but also practical for the home gardener.

For example, it is not critical that you understand mobile and immobile nutrients except that such an understanding helps you decide when foliar fertilization makes sense and why foliar feeding of calcium to prevent blossom end rot in tomatoes does not work. So I decided to include this in the book.

At other times I just included things because they are just cool to know about plants. Did you know plant roots excrete chemicals to attract beneficial microbes which then ward off root pathogens? More about this later.

To complete the book, I also had to leave out a bunch of interesting stuff and in many cases I had to take a complex topic and simplify it. I hope that the book gives you a good grounding of plant science which will allow you to find other more detailed resources as your knowledge and interests expand.

Organization of the Book

I've dissected plants into logical components including roots, stems, leaves and flowers, and each of these is discussed in individual chapters. This provides a basic background that is then used to discuss the whole plant as a single organism.

The focus of most of the book is on herbaceous plants such as grasses and perennials but a lot of the discussion also applies to woody plants. I have also added a special chapter to discuss topics that are specific to trees and shrubs. The book ends with a couple of chapters on propagation that will be very useful to gardeners and will provide better insight into specific topics that are not covered in other sections.

Numerous sidebars have been added throughout the book to discuss garden myths, which is a particular passion of mine. My blog, called gardenmyths.com, has had over 14 million visitors and discusses hundreds more garden myths.

Terms Used in This Book

Science is very precise about the terms it uses, but these are not always used in the same way by the general public, which leads to misunderstandings. One of my challenges is to use terms that are both useful to the gardener and still reflect the accuracy of the science. To ensure that we are all on the same page, it is important that we agree on some basic definitions.

Organic Matter

Organic matter is essentially dead organisms. These could be dead plants or animals, which have reached a certain degree of decomposition. Common forms of organic matter include compost, leaf mold and humus. This type of material and its role in soil is fully explored in my other book, *Soil Science for Gardeners*.

Fertilizer

The term fertilizer can have many definitions. Gardeners often think that the term only refers to synthetic chemical fertilizers, but that is not a correct usage since there are many organic fertilizers that are not synthetic.

Many jurisdictions use a legal definition for fertilizer that requires that the product contains nitrogen, phosphorus and potassium, and that the amounts of these nutrients are labeled on the package. By this definition, something like Epsom salts would not be a fertilizer even though it provides plants with magnesium. Its NPK value would be 0-0-0, which is not a fertilizer.

I will use the term fertilizer in a more general way to describe any material that is added to soil with the primary purpose of supplying at least one plant nutrient. I will also use the term synthetic fertilizer to refer to man-made chemical products and organic fertilizer for natural products.

Nutrients

When talking about plants, nutrients are the basic minerals and nonminerals they require, including things like nitrogen, potassium, phosphorus, calcium, magnesium, etc.

Herbaceous

The term herbaceous refers to plants that are herbs in the botanical sense and not in the culinary sense. It is also not used exclusively to refer to perennials. A herbaceous plant is any plant that does not form woody structures and includes grasses, perennials and bulbs. They can be annuals, biennials or perennials.

1
Plant Basics

The initial chapters of this book look at different parts of a plant such as roots, stems and leaves, but it is important to understand that none of these parts function in isolation. They are all connected to one another and although they look very different, they also have a lot of similarities. Some of these common elements are discussed in this chapter to provide a foundation for the rest of the book.

Cells

All parts of a plant are made up of cells, and unlike animal cells they have a rigid cell wall. They tend to be square in shape and the cell wall is made up mostly of cellulose, a strong polymer that adds rigidity and strength to the plant. Though cellulose is very resilient, it's also quite flexible and readily takes in water. Paper, including paper towels and toilet paper, is primarily made up of cellulose.

Cells are somewhat similar to boxes and plants can be visualized as stacks of boxes. The rigid cell walls are strong enough to withstand internal pressures and some level of freezing.

The cell wall is not completely solid. Small channels go through the wall and allow water, minerals, sugars and proteins to flow into and out of the cell, allowing material to flow throughout the plant. Water and nutrients flow from roots to the top of the plant, and sugar moves from leaves down the plant to the roots. In this way all cells are connected to one another.

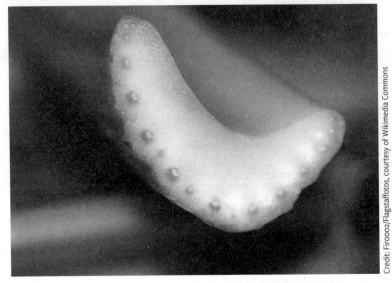

Visible vascular bundles on a celery stem.

Xylem and Phloem

The xylem and phloem are as vital to plant life as the heart and circulatory system are to animal life. In some ways, both systems play a similar role.

The xylem and phloem are two distinct organs in a plant but they usually occur together in something called the vascular bundle. Note that the term vascular is also used to refer to blood vessels, namely tubes carrying blood.

The vascular bundle is like a super-fast highway where the passengers are various substances needed for plant growth. You can easily see the vascular bundles as small dots in a piece of celery.

Think of the xylem and phloem as hollow tubes, not unlike drinking straws, running through the plant. They are more complicated than this, but the hollow tube analogy is fairly accurate.

Once water or cellular liquid enters the tubes it travels either up or down the tube. I use a simple mnemonic to remember their function. Water and xylem start with letters in the same part of the alphabet and so the xylem transports water, and water always moves up the plant, from roots to leaves.

Xylem

The xylem is responsible for transporting water from the roots to the rest of the plant. Water and dissolved minerals pass through the outer membrane of roots and then flow to the center of the root. There it enters the xylem and starts the journey up the plant.

As water is lost in the leaves it creates a drop in pressure inside the xylem that causes more water to move up the tube.

This system only supports one-way traffic. Spraying water on the leaves does result in a bit being absorbed by the leaves but this water is not moved around the plant. When a plant needs water, it must be absorbed by the roots.

Phloem

The phloem transports carbohydrates, minerals, amino acids, hormones and other chemicals produced by the plant to other areas.

Two-way traffic occurs in the phloem, so molecules in any part of the plant can flow to any other part, using phloem tubes. This system is quite complex and the plant does control the movement of molecules. Some are more easily moved around than others. For example calcium is considered to be "immobile" and it does not easily move around the plant. Other molecules like sugar are "mobile" and are easily moved to any part of the plant.

Photosynthesis

Most of you are familiar with photosynthesis. It is the process plants use to convert carbon dioxide and light into sugars and oxygen. The sugars provide the food energy plants need to grow, and the released oxygen helps us to breathe. This is obviously important for the plant, but it is also critical to life on earth.

Consider this: there were no animals on earth until plants started to grow. Initially they were simple plants like algae, but they soon developed into more sophisticated organisms. You might think that the production of oxygen was the key to animal life and it certainly was important, but plants play a much more important role and that has to do with energy.

All living organisms need energy to live. It takes energy to grow, to reproduce and to move around. It even takes energy to digest food.

Animals can't make their own energy, but plants can. Plants are the source of all our food energy and we get it by either eating plants directly, or by eating other animals that eat plants.

Where do plants get this energy? From the sun, through photosynthesis. Plants capture sunlight and create energy-storing molecules called sugars. Those sugars are then used to keep both themselves and all animals alive. Without photosynthesis, there would be very little life on earth.

A simple formula for the reaction that occurs can be written as:

$$CO_2 + H_2O \rightarrow C_6H_{12}O_6 + O_2$$

Carbon dioxide is combined with water to produce sugar and oxygen.

> ### Plant Myth: Plants Raise the Oxygen Level in Homes
>
> Plants do produce oxygen during photosynthesis and this has led to the belief that they will increase the oxygen level in the home, but that is a myth. The low levels of light in a home reduce photosynthesis which results in less oxygen being produced than outside.
>
> If you had enough plants in a room to use up all of the CO_2 and convert it to oxygen, the oxygen levels would increase from 20.95% to 21%. This increase is difficult to detect and would have no effect on humans. Keep in mind that this increase is the maximum increase possible and assumes plants would use all the CO_2 available. In real life, the increase is much less.
>
> It is also important to remember that plants also respire and produce CO_2, negating in part the oxygen they produce. If you are concerned about this CO_2 and remove plants from a bedroom at night, don't bother. The amount is tiny compared with what humans produce.

Water is collected by the roots and transported to the stems and leaves. Openings in leaves and stems called stomata allow carbon dioxide to enter. Sun is absorbed through the leaves. Chloroplasts, which are special organs in green tissues, combine these ingredients and produce sugars. A waste product of this reaction is oxygen, some of which is used by the plant, but most of it is expelled through the stomata.

What Happens at Night?
What happens to photosynthesis when the sun goes down? Almost immediately, photosynthesis stops. This means that plants need to produce enough sugars during daylight hours to support normal metabolism both during the day and all night long.

During the night, plants are still absorbing water and nutrients at the roots, and growth in various parts of the plant continues. Flowers are forming or being pollinated and leaves are producing various natural pesticides to ward off insect predation. All these processes require energy.

Photosynthesis is efficient enough to provide all the energy needed during the day and night as well as produce some excess that is stored for a rainy day, and I mean that literally.

Sugars, Carbohydrates and Energy
The terms sugars, carbohydrates and energy tend to be used interchangeably in gardening circles. Photosynthesis produces mostly glucose, a sugar, but it also produces other sugars. These sugars are simple carbohydrates and they are used as building blocks to create larger carbohydrates like starch and many other compounds found in plants.

Almost all reactions in a plant require energy and sugars provide that energy.

Each of the above terms are different but can be and are used to refer to something very similar. The glucose produced in photosynthesis can be called a sugar, a carbohydrate or even energy.

A more general term that is used by scientists is "photosynthates"—the compounds that are produced by photosynthesis. This term includes the sugars as well as many other compounds produced in the overall process.

In this book I have mostly used the term sugars and energy to refer to all of these.

Photosynthesis and Climate Change

Once you understand photosynthesis you can start to appreciate how important it is to climate change. We add CO_2 into the atmosphere and plants remove it. At present we are adding much more than plants can remove which increases the level in the air. By adding more plants to earth, we will reduce the CO_2 levels in the air and that is my excuse for buying more plants.

Most of the CO_2 absorbed by plants ends up in the soil and in woody plants. Two-hundred-year-old trees are a great storage system for CO_2. On the other hand, removing old growth forests not only stops the reduction of CO_2 by the living trees, but as the wood rots it releases the CO_2 back into the atmosphere through a process called respiration.

Herbaceous plants are also important for capturing CO_2 and that carbon ends up in soil from either root exudates (chemicals given off by roots) or decomposed plants.

ATP and the Energy Cycle

One of the most important molecules in the plant is something called adenosine triphosphate, or ATP for short. This is a relatively simple molecule that can be thought of as a rechargeable battery. Not much happens inside a plant, or even your body for that matter, without energy. ATP is that energy source.

ATP will combine with water to form another compound called ADP, as well as forming phosphate and free energy. That energy is used by thousands of other reactions to build all of the chemicals found in a plant. For example, it is used to take simple sugar molecules and build them up into large starch molecules. It is also used in the root hairs to actively move nutrient molecules across the root membrane.

$$ATP + H_2O \rightarrow ADP + phosphate + free\ energy$$

Think of ADP as a discharged battery. It is now time to recharge this battery so it can be used again and that happens during photosynthesis. The light from the sun is used to recharge ADP into an ATP molecule. The ATP battery is now ready to be used somewhere else in the plant to carry out reactions.

$$ADP + phosphate + sun\ energy \rightarrow ATP + H_2O$$

Note that phosphate plays a big role in this. It is one of the main ways in which plants use this soil nutrient.

The discharged ADP battery can also be charged through a process called respiration.

Respiration

Respiration is a process that happens in both animals and plants. An energy source, the sugars, are reacted with oxygen to produce CO_2, water and free energy. The free energy is stored in ATP.

$$C_6H_{12}O_6 + O_2 \rightarrow CO_2 + H_2O + ATP\ (energy)$$

The ATP molecule is then used as described above to carry out all the other processes taking place in a plant.

Respiration takes place in every part of the plant, including roots, stems, flowers and even in fruits and seeds. The process is not light dependent and happens 24/7. Luckily for plants, the photosynthesis process is more efficient than respiration. A well-grown plant can produce all the energy a plant needs, during daylight hours.

Respiration requires oxygen which is easy to come by in both stems and leaves which have stomata openings that allow oxygen to enter. But what about the roots? Respiration is vital for roots and the only way for them to get oxygen is from the soil. They need to breathe oxygen just like we do.

Ideal soil is about 25% air and this provides the oxygen needed by roots. This value is lower in compacted soil or if the soil is waterlogged. In both of these cases, roots can suffocate and die.

Meristematic Cells

Plants are quite different from animals in a number of respects but one that is most important is the plant's ability to keep growing. You can cut your lawn and it grows back. You can take a stem cutting and grow a complete plant. Slice a hosta in half and you have two plants. Cut down a mature tree and in no time at all you have suckers growing that will eventually grow into new trees. Try doing that to animals!

This ability to regenerate and grow is a major characteristic of plants and depends on meristematic cells.

Before we get into that, it is important to understand differentiated and undifferentiated cells. An undifferentiated cell looks like a blob. It is very simple and does not look much like a plant cell. It really can't do much yet. It is a special cell with only one function: to replicate. It divides to make more of itself and it can do this very rapidly.

Plants keep making more undifferentiated cells so that they always have a ready supply of them.

Differentiated cells are ones with a specific defined function. They might be root cells, or leaf cells or any other cell needed by the plant. They take on the physical characteristics needed as well as the biological and chemical functions required. They are mature cells ready to do whatever they were differentiated to do.

Meristematic cells are undifferentiated cells and are similar to animal stem cells which have an analogous behavior and function. They are a pool of cells ready to differentiate into whatever form the plant needs. In the root area, they might become root hair cells, but the same meristematic cell in a bud can become a leaf cell or a flower cell.

The plant keeps meristematic cells in various key locations so that they are ready to develop as needed. Primary spots for these cells include the tip of roots, the tip of shoots and in dormant buds.

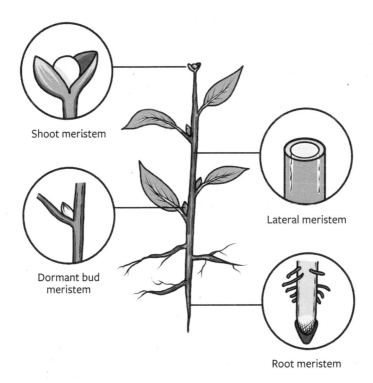

Location of meristematic tissue.

As the shoot grows, meristem tissue is also produced at all side branches in something called dormant buds. Some of these dormant buds will become new shoots or flowers the following spring, but many remain dormant for the life of the plant, or until the plant needs them. If the plant is damaged these dormant buds become active to allow the plant to grow. All the suckers produced after cutting a tree are dormant buds that have been activated by the damage.

Stems of woody plants, like trees and shrubs, have a special lateral meristem that allows the stem to grow in width.

Most growth in plants involves meristematic cells.

Unlike animals, plant growth is essentially indeterminate—both roots and shoots continue to grow throughout the life of the plant. It can slow down as the plant gets larger, and it is controlled by environmental factors, but plants never stop growing.

Classification of Plants

Plants can be divided into four main types based on how they grow and how they reproduce.

- Mosses and liverworts: simple plants that do not contain a xylem or phloem.
- Ferns: have a xylem and phloem but they do not make seeds.
- Gymnosperms: includes the evergreens and have needles and cones.
- Angiosperms: flowering plants and include most perennials and deciduous trees.

Monocots and Dicots

The angiosperms can be further separated into two distinct groups called monocots and dicots. At a high level, both groups look fairly similar but when you drill down and look at how their internal systems function you will see differences.

Monocots have a single cotyledon, flower parts are in multiples of three, leaf veins are parallel and they don't have secondary growth. Grains, grasses, lilies, daffodils and bamboo are all monocots.

Dicots have two cotyledons, hence the name dicotyledon. Flowers have four or five parts, stem vascular bundles form a ring and they tend to have secondary growth. This category includes many of the popular garden plants including peas, sunflowers, daisies, mint, marigolds, tomatoes and oaks.

In order to keep things simple, I will talk mostly about dicots and will only mention the monocots in a couple of spots where the difference impacts gardening.

2

Roots

Roots are not very well understood by gardeners mostly because they hardly ever see them. They form a vital part of a plant and root health should be a priority for gardeners; it starts with buying the plant.

Plants are purchased based on what is showing above the pot, but it is a good idea for buyers to turn over the pot and knock out the plant so that you can have a close look at the roots. You might feel awkward doing this in a nursery but no one has ever said anything to me and I do it all the time.

In most cases the roots should be white and firm. You should see some new root tips growing around the outside of the pot, but the outer surface should not be covered with circulating roots because that indicates a plant that was not repotted correctly.

The other thing to look for is brown, thin, squishy roots which indicates rotting roots, probably due to overwatering. Don't buy such plants even if the top growth looks good. When a plant dies, the roots normally go first.

One of the reasons gardeners don't understand roots very well is that once they are planted in the garden, they never see what roots are doing. You don't see root growth slowing down in mid-summer because temperatures are too high, or that they do much of their growing in fall when it is cooler. You also can't see the damage caused by daily watering or by drought conditions. At best we infer plant health by looking at top growth but root decline is normally a slow process and it can take weeks or even months to show above ground.

The more you know about roots, the better you will be able to take care of your plants.

Root Basics

Roots are plant structures that typically grow underground to stabilize the plant and obtain resources from the soil. The basic structure of a root consists of the epidermis, root cap, root hairs, cortex and vascular bundle. Root systems are highly productive and can be many times larger than the top growth.

Roots are critical to the survival of almost all plants. They anchor plants in place and obtain water and minerals from the soil.

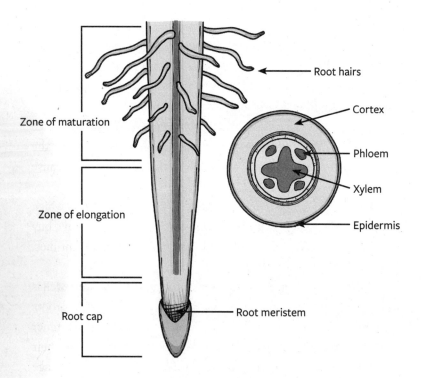

Parts of a root.

Roots help plants survive winter and other harsh environmental conditions by storing water, nutrients and carbohydrates. They can also contribute to asexual reproduction by producing root suckers or root sprouts which can become independent plants.

Epidermis

The epidermis is an outer layer of cells that surround the root. It functions as the skin of the root and allows water and nutrients to travel from the outside to the inside of the root. It also controls the movement of plant-produced chemicals, collectively called "exudates," to travel out of the root.

The epidermis also allows oxygen to move into the root, which is critical for respiration.

Cortex and Vascular Bundle

In simple terms the inside of a root can be divided into the cortex and the vascular bundle (xylem and phloem). The primary function of the cortex is to conduct molecules between the epidermis and the xylem and phloem. In thicker roots it is also important for storing carbohydrates produced by the leaves. A mature carrot root is mostly a thick cortex full of these carbohydrates.

The xylem moves water and nutrients from the soil to all other parts of the plant, while the phloem moves minerals, proteins, sugars and other photosynthates from the upper plant to the roots.

Root Hairs

Gardeners talk about roots absorbing water and nutrients but it is actually the root hairs that do most of this. Root hairs look like very fine white hairs coming out of the roots. They tend to be too small to see them in most soil, but they are very visible on newly germinated seedlings.

Root hairs are an extension of the root epidermis. Most are formed near the root tips and they only last a couple of weeks. After that, the root aborts them and grows new ones closer to the root

Tip of a root showing the root cap and the root hairs.

tip. A plant's ability to replace them quickly is important for its survival. For example, if things get too dry, the plant aborts the root hairs, but keeps the roots. When enough moisture returns the plant quickly grows new root hairs to make use of the water.

Why does a plant make root hairs? Why not just rely on the root? By growing thin root hairs the plant has dramatically increased its root surface area. More surface area means more access to water and nutrients. Root hairs are hundreds of times more efficient at collecting water and nutrients than roots alone.

Root Cap

The root cap consists of several layers of cells in a dome formation that forms a thimble-like covering over the very tip of the root. The root cap protects the sensitive new tissues of the root tip from damage and infection as the root elongates and pushes through the soil. It functions similarly to the cap on a pen protecting the tender tip. The outer layer of the root cap is made up of dead cells that get rubbed off by soil friction, only to be replaced by new cells.

How Do Roots Get Longer?

Roots grow by adding length to their tips. This involves four key parts within the root tip: the root cap, the root (apical) meristem, the zone of elongation and the zone of maturation.

It's actually very difficult for roots to push their way through soil, so they attempt to use existing openings made by animals like earthworms and by decomposed roots. If there are no available openings, the root cap will secrete a sugar-based lubrication called mucilage.

Compact soil lacks openings between soil grains, making it difficult for plants to grow roots. Soilless potting media and sandy soil are very porous with lots of air spaces which explains why roots grow so well in them.

The root meristem is located right behind the root cap where it actively produces new undifferentiated cells. Some of these cells will differentiate into new root cap cells and others into new root cells. This active addition of new cells pushes the root cap forward, lengthening the root.

The zone of elongation is immediately behind the meristematic zone. This is where newly formed root cells grow to their full size, pushing the root cap even further. As these cells mature to take on their final functions they become part of the zone of maturation. It is at this point that one can more easily recognize different types of cells: epidermis, xylem, phloem, etc.

At the same time root cap cells are also differentiating and maturing but their life is short. Their main function is to form a hard root cap, but even so, they don't last long before friction scrapes them off.

Types of Roots

There are four main types of roots: fibrous roots, taproots, tuberous roots and adventitious roots. Plants may use a combination of different types to maximize growth, survival and reproduction.

Fibrous roots are what typically come to mind when you picture plant roots. They consist of thin, branching roots that form a dense network close to the soil surface. All of the roots are of a similar length and girth. Fibrous root systems have a high surface area, allowing the plant to take up a lot of water quickly and they are very efficient at stabilizing the plant and preventing soil erosion.

Taproots are characterized by a thick, primary root that grows deep into the soil. Think carrots and dandelions. Thinner, secondary roots develop along the side of the taproot. This is a common root system used by plants that have evolved drought tolerance. They also act as a good food reserve and many are able to regrow if the top part of the plant is removed which explains why dandelions regrow after weeding.

Plant Myth: Feeder Roots Are Located Under the Dripline

The common drawing of a tree shows a globe above ground and a similar globe below ground. The dripline is the outer edge of the aboveground growth. It is believed that this is where one finds the ends of the roots but this is almost never the case.

The root ball looks more like a flat dish than a globe, with most of the roots near the surface of the soil. Secondly, the edge of the root system is two to five times wider than the dripline which means most feeder roots are also outside of the dripline. Water and fertilizer should be applied outside of the dripline, not at the dripline.

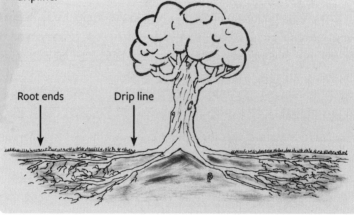

A tuberous root, or root tuber, is an enlarged, fleshy root that is used primarily for food storage. Plants with this type of root system usually produce fibrous roots beneath the tuber. Examples of tuberous roots include dahlias, peonies and sweet potatoes.

You might think that potatoes are also root tubers but they are really stem tubers, as are tuberous begonias. It can be difficult to tell them apart.

Adventitious roots are roots that grow from stems or leaves, rather than the existing root system. They take on different forms and roles. Climbing plants, like ivy, produce adventitious roots to cling onto surfaces. Plants with runners, like strawberries, develop adventitious roots along the stems to establish new plants. Adventitious roots are a common response to stress, injury or wounding. You may have noticed that some succulent plants produce roots from the stem. These are aerial (aboveground) adventitious roots that are produced because of lack of water.

Adventitious roots that remain above ground do not contribute to the collection of water and nutrients.

Effect of Gravity

Plants are constantly experiencing gravity and respond to it through a phenomenon called gravitropism—the response of a plant to gravitational pull. The roots grow downwards with gravity while the stems resist gravity to grow upwards to maximize light absorption.

Downwards root growth is maintained even if a plant is tipped on its side, or if it is planted upside down. Common advice is to plant bulbs the right way up, but they don't really care. Even if planted upside down, roots grow down and shoots grow up.

The roots of seedlings are strongly affected by gravity and they head straight down. As the plant matures, gravity plays less of a role and water resources become a more dominant factor. You see this clearly when you build a raised garden over tree roots. In no time at all the tree roots have filled the raised bed. The reason for this is that the soil in the bed has good air, water and nutrients, so roots ignore gravity and grow up.

Many trees form an initial taproot in part due to strong gravitational effects, but over time the tree grows more and more surface roots to take advantage of the resources there.

Cut Roots and Side Roots

The growth of roots discussed so far explains how roots get longer, but how do they make side branches and what happens if you cut the root tips while moving the plant? In both cases there is no root tip to continue growing.

Roots have a special ability to handle this. They are able to form new meristematic tissue anywhere along their length. Once formed, it produces new cells that then differentiate into a new root cap which grows into a new root.

If a plant wants new side roots, it simply grows a new root cap.

> ### Plant Myth: Circulating Roots Continue to Circulate
>
> It is believed by many that circulating roots in a pot will continue to grow in a circular fashion after being planted, but that's a myth. The reason these roots are circulating is not due to any internal need to circulate. They are trying to grow in all directions away from the center of the plant, but the plastic prevents them from doing that. The hard plastic causes circulating roots.
>
> When these roots are released from the plastic they will start to grow out from the crown of the plant. This works fine for herbaceous plants but woody plants have a special problem. The roots of woody plants continue to expand in width and get thicker every year. They start to strangle themselves. As they run out of space they can also expand above ground and around the trunk of the tree. When this happens they strangle the trunk as well, eventually cutting off the flow in the phloem. Once that happens the tree is likely to die.
>
> Cutting the roots before planting helps them grow in new directions and in woody plants it can eliminate damage to the trunk.

It does the same thing if the tip of a root gets cut off. New side roots grow where needed.

This ability is important to gardeners. Planting from a pot is very likely to do some root damage. Moving an established plant does even more damage and up to 70% of the root system could be lost. This is significant harm to the plant, but it can recover by forming lots of new side roots.

Many purchased potted plants have been in the pot far too long which results in circulating roots. When you plant these in the ground, it is a good idea to cut through the outer circulating roots. Make four vertical cuts, each about half an inch deep. This seems very drastic, but it results in many new root tips growing into the surrounding soil.

How does severe root damage affect the whole plant? The plant can regenerate new roots but this does take time and while it is going on, the roots can't provide enough water and nutrients for the aboveground part of the plant. With small damage you will see dropping leaves and maybe even some browning of the leaves as leaf tissue dies.

In severe cases, there are not enough resources to support both leaves and root growth and the plant usually prioritizes roots resulting in leaf drop. If the damage is done early in the year, new leaves will be grown once a better root system is established. In the long run this is better for the plant.

Now that you understand this, you also understand why the common advice to water a new plant with droopy leaves makes no sense. It can't absorb excess water because it lacks a complete root system. Water only when the soil starts to get dry.

Conditions That Affect Root Growth

Roots are dynamic and opportunistic. They will grow where they find the best conditions.

Imagine a newly planted perennial. It has a good root ball with roots on all sides. We expect the roots to start growing in all directions, like the spokes of a wheel, but that may not happen. If you water more on one side than the other, more roots grow on the wet side than the dry side. If one side gets more fertilizer roots grow mostly on the nutritious side.

The amount of root growth is also affected by top growth. A plant in a shady spot makes fewer leaves and does less photosynthesis, resulting in lower demands on the root system. The plant responds by growing fewer roots.

What conditions do roots want?

- 25% air
- 25% water
- Easy access to nutrients, provided the levels do not become toxic
- pH between 5.5 and 7.0
- Temperature between 40°F (4°C) and 90°F (32°C)

> ### Plant Myth: Transplant Solutions Grow Better Roots
>
> A range of products are available for helping plants get over the shock of being transplanted and they go under various names including transplanters, root stimulator and root booster. These products are essentially fertilizers that are added at the time of transplanting. They usually have a NPK with a high middle number which is the phosphate level.
>
> Don't use transplant solutions.
>
> 1. It is unlikely that your soil is deficient in phosphate and so you don't need to add more.
> 2. New transplants should not be fertilized.
> 3. High phosphate does not stimulate root growth. Roots need all of the nutrients in the same ratio as other parts of the plant, namely an NPK of 3-1-2.
>
> Don't confuse these products with rooting hormones which are used for making roots on cuttings. They should not be used on transplants.

A gardener can control some of these, but not all. Soil compaction squeezes out air making it harder for roots to grow—so don't walk on soil.

Watering forces air out of the soil as water replaces the air. Roots can handle this provided you are not watering every day. Many gardeners water way too often and starve roots of much needed oxygen.

Mulching keeps soil warmer in winter and cooler in summer.

Root Competition

How does root competition affect plant growth?

If you consider only the available space in soil you find that there is very little competition. The volume of soil is much greater than the volume of roots even in a crowded garden. In good soil competing plants don't have trouble finding space to grow roots.

There is however competition for resources, namely water and nutrients.

Why can't you grow things under evergreen trees? It is not because the soil is too acidic, which is a very common myth. The reason is that evergreen trees make many fine roots that are very good at sucking up water. Other plants trying to grow there have a very dry environment which prevents proper growth.

There is also the allelopathic effect which is discussed in the root section called "Allelochemicals."

Competition between roots has a minor effect on plant growth compared to the effect of light competition above ground.

Root Growth in Winter

Root activity is periodic. They grow mostly in spring but also in fall once temperatures get lower. That does not mean the whole root system grows at one time. Root growth in any particular spot depends on environmental and plant conditions.

They can survive down to about 20°F (–6°C) which is much higher than the minimum survival temperature of tree branches. The main reason for this difference is that the aboveground growths are fully hardened off with most metabolism stopped. The roots on the other hand do not get hardened off to the same degree, so that they are ready to grow as soon as conditions improve.

Roots start to grow when soil is a few degrees above freezing. They will take advantage of a winter thaw and start to grow, only to shut down again if things get cold.

The ability to take advantage of warm spells is critical for trees, especially evergreens that need to collect water in winter so they don't desiccate. It also allows roots to spread, ready to take advantage of spring rains.

Absorption of Water and Nutrients

The root epidermis is composed of cellulose, a spongy material that absorbs water. Once the water is inside the root, it travels to the xylem where it is carried to the rest of the plant. Most of this absorption happens at the root hairs.

The absorption of nutrients is a bit more complicated. Most plant nutrients are ions which have a positive or negative charge. Calcium, magnesium, iron and manganese are examples of positively charged ions. The negative ones include nitrate, sulfate and phosphate. This charge is important because it affects how they behave in soil.

Clay and organic matter are negatively charged and the positive ions stick to them, much like the opposite poles of a magnet attract each other. This makes it difficult for roots to absorb these ions. To solve this problem they use a technique called a proton pump, where they pump positively charged hydrogen ions into the water surrounding the root. The hydrogen ions also stick to clay and they displace the other positive ions, freeing them so roots can absorb them.

The negative ions are less attracted to clay and organic matter, and are free to float around in the soil water making it easier for roots to absorb them. This might sound great for plants but it does cause some issues.

These negative ions, especially nitrate, move too easily with the water. A good rain or a heavy-handed gardener quickly washes excess nitrogen down below the root zone. This partially explains why nitrogen is the nutrient that is most likely in short supply.

When fertilizer, compost or manure is added to the top of the soil, water quickly washes the nitrogen to the roots. Leafy vegetables like lettuce, have become very good at trapping the nitrogen and they can absorb too much. Early lettuce can have nitrate levels that are unhealthy for eating.

Osmosis and Fertilizer Burn

Water moves in and out of roots by a process called osmosis. This is also the way water moves between cells throughout the whole plant.

Water moves from areas of low solute concentration to areas of high concentration. *Solute* is a generic term for most compounds including nutrient ions, sugars, amino acids, proteins, etc. The root maintains a high level of solutes inside the epidermis while the level of these outside the root is normally low. This causes water to flow into the root by osmosis.

If you fertilize your lawn too much, it will burn the grass which shows up as brown grass leaves. Trees and shrubs end up developing brown spots at the tips of leaves. It is also a common problem with houseplants. Most consider this to be a form of toxicity. The nitrogen levels got too high and became toxic, but that is not what happens.

If too much fertilizer is added to soil, the level of nutrients goes up. This is especially true for negative ions like nitrogen which don't bind to clay. As the solute concentration outside roots goes up, it can become higher than the solute concentration inside the roots. When this happens, osmosis causes water to flow out of the roots and the plant is starved of water. Leaves go brown because of a lack of water, not a toxic effect.

Active Transport

Unlike water, nutrients do not easily flow into roots for two reasons. First, they are charged ions and such molecules can't move across cell membranes. The second reason is that the concentration of nutrients outside the root is lower than inside of the root. It is akin to swimming upstream—it doesn't work well.

To overcome these problems plants use a process called active transport to move ions across cell membranes. This is critical in roots, but the same process is used to move nutrients throughout the plant from one cell to another.

Proteins on the outer surface of the cell wall grab specific nutrients as they move past the root hair. They then pass the captured molecule to another protein on the inside of the cell wall, where it is released into the root. This description is a simplified version of a much more complex process which scientists are still trying to fully understand.

The key point here is that roots are able to control which molecules enter the plant, at least to some extent. Large molecules like proteins and carbohydrates might not be able to enter. Smaller molecules like amino acids might be absorbed if there is an active transport mechanism for them.

Why is this important to gardeners? Many manufacturers of products list all kinds of molecules in their product that are "useful to plants." In many cases we don't know if plants can actually absorb them. Scientists are still trying to figure out what plants can and cannot absorb. Just because a product contains compounds that could be useful to plants, does not mean plants can actually absorb them and if they can't, they are not useful to plants. Many of these products make claims that are simply not true.

pH and Nutrient Availability

Soil pH can have a significant effect on plant growth and that is why so much has been written about it.

pH measures the amount of hydrogen ions in a liquid. The value is given as a number between 0 and 14, with a pH of 7 being neutral. Anything above 7 is alkaline (low level of hydrogen ions) and below 7 it is acidic (high level of hydrogen ions). Most plants grow best at a pH between 5.5 and 7.0, but some plants prefer a value above this and others below it.

You have probably seen suggestions that the ideal pH range for plants is 6 to 7. This is true for mineral soils which are made from rock, but for organic soils (peat and marsh bogs) a better range is 5.5 to 6.

Given all of the talk about pH and plant growth you would think that pH is important but the reality is that plants are not directly affected by it. The issue with pH is that it affects the nutrient levels in the soil solution which in turn affects plant growth. The chemistry gets a bit complicated so I'll just provide a few examples to illustrate the issue.

Assume a soil has plenty of nutrients. If this soil is acidic, phosphorus reacts with aluminum and precipitates (becomes a solid), reducing the amount available to plants which can't get enough of either one to grow properly. There is still a lot of phosphorus and aluminum in the soil, but it's in a form that plants can't use.

Plant Myth: Is Soil pH Important?

People talk about "soil pH" but what they are actually referring to is the pH of the water surrounding soil particles—the soil solution. A measured value is the average from the total soil sample. Specific locations within soil can be quite different. A spot with high organic matter and a lot of bacterial activity will have a different pH than an inch away where there is less organic matter. The pH right next to a clay particle can be as much as one pH unit different from the water that bathes the clay.

The rhizosphere, the area right next to plant roots, can have a very different pH than the soil solution.

Is soil pH important? pH is certainly important, but gardeners should only use it as a rough guide for selecting plants. Don't be too concerned about your soil pH and don't try to make it perfect. Soil pH is very difficult to change especially if it is on the high side.

If you want to learn more about soil have a look at my book *Soil Science for Gardeners*.

On the other hand, if this same soil is alkaline, calcium reacts with iron and reduces the amount of iron available to plants, causing them to show interveinal chlorosis. The problem is not a lack of iron, but a lack of available iron, due to pH.

Adding more aluminum or phosphate in acidic soil, or more iron in alkaline soil does not solve the problem, because the added amount just becomes unavailable to plants. It is a pH problem, not a nutrient problem.

Most plant nutrients are affected by pH. As a general statement, nutrients are more available at a neutral pH and become less available at extreme pH values. The chart shows how the availability of various nutrients changes with varying pH. Keep in mind that the chart shows relative amounts. It can't be used to calculate the amounts available at any given pH.

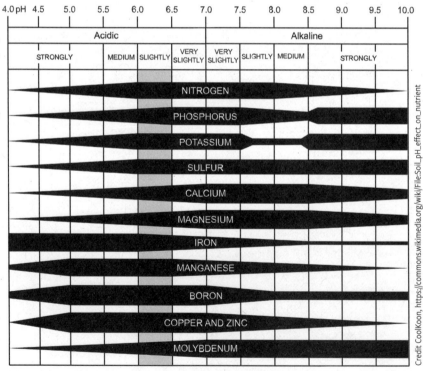

pH affects nutrient availability.

pH and Nutrient Toxicity

Plants need certain nutrients but these same nutrients can become toxic to the plant if they exist in concentrations that are too high. At a low pH, aluminum and manganese can reach toxic levels.

Aluminum toxicity is not usually a problem in mineral soils that have a low amount of clay or in peat based soils. Some people recommend aluminum sulfate as a soil conditioner, but due to possible toxicity, it should not be used on clay soils.

Roots and Microbes

We think of plant roots as being isolated in the soil but that is far from the truth. They form a number of associations with microbes.

Mycorrhizal Fungi

Mycorrhizal fungi are a special group of fungi that form a very close symbiotic association with about 95% of all plants. The fungi attach themselves to roots in such a way that they can transfer compounds back and forth. Both the plant and the fungi benefit from this arrangement. The fungi receive sugars and nitrogen containing compounds in exchange for water and mineral nutrients. The average plant shunts about 15% of its food to the fungi but in some plants it can be as high as 30%.

Think of the fungi as having very fine filaments like the strands in a ball of cotton. These filaments, called hyphae, are much thinner and longer than roots and have a much easier time reaching cavities in the soil that are too small for roots. Fungi can extend the effective root system of a plant by a factor of 1,000, greatly increasing its ability to access nutrients.

Phosphorus levels are always low in the soil solution, but fungi are able to pick up significant amounts due to their large surface area. This is one of the most important nutrients they provide to plants. It should be noted that large amounts of phosphorus in soil, from too much fertilizer, will inhibit mycorrhizal growth.

These fungi also protect plant roots from disease and can absorb toxins before they reach the roots.

Natural soil contains lots of different species, but cultivation, removing top soil and compaction reduce diversity, making it more difficult for the fungi to grow.

Nitrogen-fixing Bacteria

Nitrogen is a critical nutrient for plants and it is the most likely nutrient that is deficient from soil. It can be added using synthetic fertilizer or organic fertilizer such as compost and manure, but there is another very important source of nitrogen; bacteria.

Specialized nitrogen-fixing bacteria are able to take the nitrogen gas in the air that plants can't use, and turn it into ammonium and nitrate ions, which plants can use. This process is called nitro-

> **Plant Myth: Purchased Mycorrhizal Fungi Are Good for Plants**
>
> Natural mycorrhizal fungi are good for plants so many companies now sell packaged fungi to gardeners. They claim that garden soil doesn't have enough fungi and adding more will improve the health of the soil, which in turn grows better plants.
>
> There is a fundamental flaw with this logic. Garden soil may have low levels of fungi—that part is true, especially in new homes where all the top soil was removed and the remaining subsoil is heavily compacted. This type of soil does not have the organic matter needed to support fungi. The problem is that adding more fungi does not change the organic levels or the degree of compaction, so the added fungi just die.
>
> If soil supports the fungi, they will already be growing there. Fungi spores are everywhere and constantly arrive in your garden by wind. You do not need to buy them. If you want more fungi, improve the soil and they will come.
>
> Still not convinced? Go to an established wooded area or meadow and collect a handful of soil. It will contain more mycorrhizal fungi than any purchased bottle.

gen fixation. These bacteria exist in two basic forms; free-living and symbiotic.

Free-living Nitrogen-fixing Bacteria

Free-living nitrogen-fixing bacteria exist throughout the soil and are not directly associated with roots. Most of them need a carbon source for energy like dead organic matter but a few types can use minerals as an energy source.

These bacteria produce an enzyme called nitrogenase which fixes nitrogen outside of their body. Oxygen inhibits this enzyme and therefore most of these bacteria live in anaerobic conditions. In natural situations these organisms are not a major contributor to the total amount of fixed nitrogen, but they can be important in special environments. For example, some species tend to live in the rhizosphere (area around the roots) of grasses and cereal crops, where they contribute a significant amount of nitrogen.

Symbiotic Nitrogen-fixing Bacteria

The symbiotic nitrogen-fixing bacteria form relationships with specific plants. The best known of these is the association of *Rhizobium* bacteria with legumes, such as clover, alfalfa, soybeans, broad beans and peas.

The plant initiates contact by releasing flavonoids through their roots. The bacteria sense these compounds and attach themselves to specialized root hairs. The legume then forms a growth, called a nodule, around the bacteria. This nodule not only protects the bacteria, but it also provides an anaerobic environment which is important for nitrogen fixation.

The plant feeds the bacteria by excreting sugar and other nutrients into the nodules. The happy bacteria then fix nitrogen and make ammonium, which is converted to nitrate as it is absorbed into the plant. You can think of these nodules as being little factories that make nitrogen for the plant and the bacteria are the workers.

This ability of legumes to have their own nitrogen source means that they are able to grow in environments that have very low

Nitrogen-fixing root nodules on wisteria (hazelnut for scale).

natural nitrogen levels, making them very competitive. But there is a cost to the plant for this nitrogen. The plant may use as much as 20% of its photosynthates to maintain the bacteria.

There are numerous species of *Rhizobium* and each one is specialized for a type of legume. For example, peas and beans are infected by different species. In order for legumes to form nodules, the bacteria must be present in the soil. If your soil does not contain the right strain, no nodules will be formed.

Gardeners solve this problem by inoculating seed with the right bacteria at the time of planting. Little packs of bacteria can be purchased from seed companies or you can buy seed that is already coated with the right bacteria. Once the bacteria is in the soil, it will survive there for several years, so even a 4-year crop rotation does not need to be inoculated each time.

How do you know if you have the right bacteria in the soil? Grow the legume and have a look at the roots in mid to early fall. You can easily see the pea-size nodules if they are there. They are most visible when plants bloom.

> ### Plant Myth: Roots Grow Towards Water
>
> The idea that roots grow towards water is a myth. Roots cannot look out several feet into the soil and sense water. They can't even sense water a few inches away.
>
> Roots do react to water right at the root tip, a process called hydrotropism. If one side of the root cap is wetter than the other side, the root cap sends a signal back to cells in the elongation zone. This triggers the cells on the dry side to elongate more, causing the root to change direction and grow towards the wet side. But this sensing of moisture only extends a fraction of an inch past the root.
>
> Why do roots grow in sewer pipes? If water carrying pipes develop a crack it is possible that a root will start growing into the crack, purely by chance. Once inside, the excess water causes more roots to develop. Before long the pipe is full of roots. But roots do not crack the pipe and don't grow towards water pipes. They only grow there if they stumble upon them and find water.

If the plant did not make nodules you either do not have the right bacteria or you have too much nitrogen. Excess fertilizer will prevent the formation of the nodules since the plant simply does not need the bacteria.

The Rhizosphere

The rhizosphere is a very special area in soil that is very different from other places in soil. Microbe populations can be a thousand times higher and it is full of chemicals that are excreted by both plants and microbes. It contains a very high level of fresh organic matter and the pH can be quite different than anywhere else in soil.

This unusual place is a thin 0.1 inch (2–3 mm) area around roots, especially near root tips.

Because of the high population of living organisms and intense competition, this area has a lower pH, higher organic level, higher

CO_2 concentration, lower nutrient content, fewer contaminants and lower water resources than the rest of the soil.

The rhizosphere is critical to plant growth and yet it's not well understood by gardeners or farmers. Even scientists are just now starting to unravel its mysteries.

In many ways soil resembles our rural and urban areas. Rural areas account for most of the soil space and accommodate few people. As you move towards the city, the population increases and things get more congested. The population density and metabolic activity is highest right downtown where the roots grow.

Root Exudates

Roots release all kinds of chemicals that as a group are referred to as root exudates. Scientific research has shown that they are produced for various reasons:

- Restrict the growth of competing roots
- Attract microbes in order to form symbiotic relationships
- Change the chemical and physical properties of the soil and soil solution
- Make nutrients more available

The exudates include sugars, carbohydrates and proteins which form a perfect food source for microbes. Sugars and carbohydrates provide the carbon and proteins provide the nitrogen—the perfect buffet for living organisms.

When a root enters a new area of soil, the first to join the party are the bacteria and fungi. Bacteria in particular show explosive growth as they feed on the exudates. This is quickly followed by their predators: nematodes and protozoa. As the microbe populations increase, they each contribute their own excrements, adding to the nutrient pool. The water in the rhizosphere becomes full of thousands of different chemicals.

Plant roots also produce amino acids, vitamins, organic acids, nucleotides, flavonoids, enzymes, glycosides, auxins and saponins. Some of these attract specific organisms that plants want to have nearby, while others are toxic to certain species. What is amazing about this is that through the use of exudates, plants have some control over the type and number of microbes that inhabit the rhizosphere. They even change populations based on seasons and environmental conditions.

The exact nature of the exudates depends on many things including genetics, age of the plant, light, water, temperature, mineral deficiencies and even reduction of leaf surfaces due to pests and diseases. Herbicides, antibiotics and fertilizer sprayed onto plants also affect the exudates and we know very little about these effects.

Producing these chemicals is a major drain on plant resources. It is estimated that some plants excrete 50% of the fixed carbon from photosynthesis through their roots. Annuals excrete 40% of their photosynthates and for other plants the typical value is around 30%.

The large number of microbes results in a lot of dead microbes, both due to short life spans and predation—microbes eat microbes. This produces high levels of nutrient ions right next to the roots. This is much more efficient than trying to find and extract nutrients out of soil.

Most of the activity in the rhizosphere occurs near the tip of the root which is also the point where most of the water and nutrients are absorbed by the plant, so it makes sense to also produce the exudates here. Cells in the root cap and root hairs are also an important component of the rhizosphere since they are short lived. Root caps shed up to 10,000 cells each day and root hairs only live a couple of weeks. Combined, these two sources add a significant amount of fresh organic matter; a perfect food source for microbes.

Dynamic Microbe Populations

The population of microbes in the rhizosphere is complex and varied. Many factors affect the type and number.

There are many beneficial interactions between microbes and plants:
- Specific exudates attract mycorrhizal fungi to the roots.
- Seeds produce different compounds before and after germination in an effort to manipulate the microbe populations in their favor.

- Legumes produce flavonoids to attract rhizobia so the plants can form nitrogen-fixing nodules.
- Special bacteria called plant growth-promoting bacteria colonize the rhizosphere and produce plant hormones (auxins and cytokinins) that increase plant growth and improve mineral uptake.
- Microbes can also produce chelates which make nutrients more available.
- Inoculation of potatoes with bacteria has shown a 17% increase in yield, probably due to the antibiotics they produce, which helps control pathogens.
- Other bacteria release enzymes that inhibit the development of nematode eggs, decreasing root damage.
- *Azospirillum* are bacteria that live around the roots of grasses and fix nitrogen.

Science is just now starting to discover and understand these associations, but it is clear that the microbe life in the rhizosphere is critical for plant growth.

Allelochemicals

Allelopathy is a biological phenomenon where an organism produces chemicals that affect the growth and survival of another organism. Allelochemicals can have beneficial or detrimental consequences. In the strict definition this applies to any type of organism, but the term is commonly used by gardeners to describe interactions between plants and most of the discussion centers around detrimental effects.

The most well-known case of allelopathy is the effect walnut trees have on the growth of other plants. It is believed by many that nothing grows under a walnut tree because it excretes a chemical called juglone.

The reality of this story is quite different. The various parts of a walnut tree don't contain any juglone. They do have another compound that turns into juglone on exposure to air, but this quickly converts to other compounds. Juglone will prevent some seedlings, notably tomatoes and other nightshades, from growing but most of the studies showing this have been done in the lab and not the real world.

Nobody has been able to demonstrate that plant roots actually take up juglone from the soil. If they don't absorb it, how can it affect plant growth? It is true that some plants don't grow as well under a walnut tree, but most mature plants grow just fine.

Other examples of allelopathy include garlic mustard (*Alliaria petiolate*) which inhibits the growth of other plants and mycorrhizal fungi. Mexican fireweed (*Bassia scoparia*), an invasive weed in central North America, reduces the growth of spring wheat.

There are many other claims of plant to plant allelopathy but most of the details of these are not well understood. The chemical effects of allelopathy reach beyond the rhizosphere, but weaken as one gets farther away from plant roots.

I suspect the effect is real, but not nearly as potent as claimed in gardening circles.

Plants Are in Control

We think of plants as being passive about getting their nutrients. Plants do grow roots in search of nutrients, but this growth is fairly random because they don't know where the nutrients are until a root stumbles onto them. This view of plants as helpless is misleading. Plants play a major role in bringing nutrients to the roots.

As discussed above, plants are actively attracting and herding the right microbe food sources to live next to their roots, but they even take an active role in extracting nutrients from soil.

Acidic exudates change the pH of the soil solution around the roots, which helps dissolve insoluble minerals in the soil, making them plant available. The acids are dissolving the rocks around the roots. The acidic condition also makes phosphate (PO_4) more available to roots.

The production of all of these exudates is very much under the control of plants. If phosphorus levels are low, they produce more acids. If iron levels are low, they produce chelates to trap iron. As nutrient levels rise, they produce less sugars because they no longer need a big microbe herd. Why feed them and encourage their growth when the plant has enough food?

Plants manipulate the rhizosphere and its populations to benefit themselves.

It is important to understand that the above sentence does not imply any sort of knowledge, thinking, planning or intelligence on the part of plants. All of this is controlled by basic chemical reactions, many of which are controlled by enzymes which have the capability to change their activity based on the presence of certain triggers. For example, at high iron levels, an enzyme may be present but inactive. When levels drop, it becomes more active, which results in the production of a special chelating exudate, making it easier to absorb iron molecules. As iron levels increase, it again shuts down activity until needed. This all happens through the magic of chemistry—not plant intelligence.

3

Stems

Stems connect roots to leaves and can take a variety of shapes and sizes. Some are short and stiff, others are long vining organs, and still others don't look like stems at all as in cactus. They can be a few inches in length or grow to hundreds of feet. Most stems grow above ground but some modified stems like bulbs, corms, tubers and rhizomes grow underground.

Herbaceous stems are green and flexible and are found on annuals, biennials and perennials. They die back to the ground at the end of the growing season, only to return again in spring. Woody stems start out being herbaceous and then go through a maturation process to become hard. This type of stem will be discussed in more detail in the chapter on woody plants.

Stems perform several functions:
- Provide structure to the plant and elevate leaves to get more light
- Transport water, nutrients and sugars between roots and leaves
- Carry out photosynthesis
- Store food for the plant

The term shoot is used to refer to a stem or to describe newly forming leaves or flowers. It's also what you say after a plant dies—shoot, I killed another one!

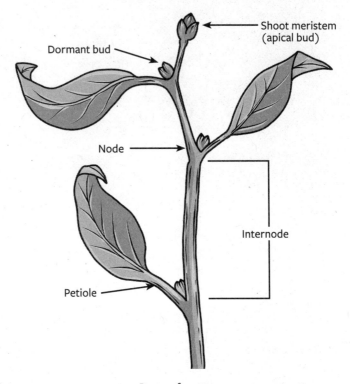

Parts of a stem.

The Outer Structure of Stems

Some stems are unbranched like the palm tree, but most garden plants have stems with side branches. The point at which the branch connects to the main stem is called a node. The stem region between two nodes is called an internode.

Nodes are critical to plants as they are the connection point of branches, leaves, aerial roots, flowers and dormant buds. Even when these structures fall off, the node is still visible as a scar.

Internal Structure of Stems

A cross section of a stem reveals its internal structure which includes the epidermis, cortex, vascular bundle (phloem and xylem) and pith.

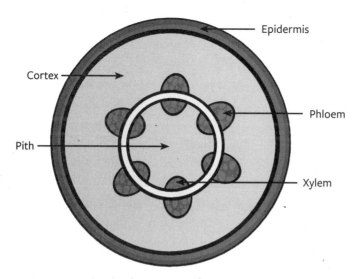

Internal structure of a stem.

The outer layer of a stem is the epidermis and it is very similar to the epidermis of a leaf. It has stomata to allow air exchange and it allows light to pass through. It's coated in a waxy material which provides protection from wind and rain as well as adding strength to the stem.

Most of the functioning parts are found inside the cortex with photosynthesis taking place just under the epidermis. The vascular bundles contain the xylem and phloem. In dicots they are arranged in a circular pattern, located between the cortex and the pith. In monocots they are distributed throughout the cortex.

The cortex also contains various fibrous cells that grow into long, stiff structures giving the stem its rigidity. The strings in celery stalks are examples of this. These fibrous cells are still flexible in the newly formed tip of the stem which is easily bent. As the stem matures, these cells become more rigid, making the stem stiffer. If you have ever harvested asparagus you have experienced this phenomenon. Newly picked spears are soft and flexible—good for eating. In a few days they become stiff and woodier as their fibers become more rigid.

In some plants, like bamboo, the whole stem becomes very fibrous and stiff, allowing the plant to grow as high as 50 feet. These herbaceous stems are so hard that they resemble wood. The so-called coconut tree is not a tree at all. It's main trunk is a stiff stem, but it still has some flexibility allowing it to bend in strong winds.

The pith is the innermost part of the stem and it can be made up of spongy material or in some cases, like goldenrod and forsythia, it's hollow. It is completely absent in monocots.

We talk about stems and roots as separate plant organs, but of course they have to be connected. The epidermis of the root, leaves and stem are really one continuous structure as are the vascular bundles and the cortex.

Trichomes

Instead of root hairs, a stem will produce extensions on its epidermis called trichomes. These can take the form of hairs, spines or scales. The fuzzy outer coating on lamb's ears (*Stachys byzantina*) is due to trichomes.

Like human body hair, trichomes protect the plant from too much sun, keeping it cooler and reducing water loss. The thorns on roses and raspberries protect against animals and in some cases they secrete chemicals that deter pests.

Buds

The stem contains two basic types of buds: terminal and lateral.

Terminal buds, also called apical buds, are found at the tip of growing stems, including the tip of side branches.

The buds located in the leaf nodes are called lateral buds or axillary buds. Flower buds are examples of these.

In dormant herbaceous plants there are no visible stems above the ground but they still contain buds. These are usually located in the crown of the plant or on other underground stem structures like tubers. If you have ever dug up a hosta or peony in fall or early spring you will have noticed them since they are quite large and pointing up. The so-called eyes on potato tubers are also dormant buds.

Dormant buds are any buds that are not actively growing. Many lateral buds are dormant, waiting for the right time to grow and even the terminal buds on woody plants are dormant in winter. Forsythia and magnolia make dormant flower buds in fall so that you can enjoy an early spring bloom.

Buds contain meristematic cells that allow the bud to grow. These cells will develop into various differentiated structures including stems, leaves and flowers.

How Stems Grow

Herbaceous plants spend the winter underground. In spring, the warming temperatures signal the plant to start growing. Dormant buds located on the underground structures start to develop. The meristematic tissue in the buds starts to reproduce and begins the process of differentiation. Some cells become more stem cells, others become leaves or flowers. As the stem gets longer some of this meristematic tissue remains at the growing tip, rising as the stem forms behind it.

Some of the meristematic cells also stay behind and become lateral buds. Depending on conditions, these may also start to grow, or the hormone levels in the plant will keep them dormant.

New stem cells start out very flexible and as they mature they become harder and special fibrous structures develop. Over time, this gives the stem rigidity. The degree to which this takes place depends very much on genetics. Some stems become very stiff and strong. Others remain quite flexible.

This process is very evident in vines like clematis and pole beans. You can easily push the top of the stem in any direction to get it growing the way you want to get it to attach itself to your trellis. If you try the same thing a couple of weeks later, you will break the stem. This is why it is important to check these plants every week or so and direct their growth.

What happens if you break the stem off? The terminal bud is now gone and this allows lower lateral buds to start their growth process. These become the plant's new terminal buds. Gardeners use this to their advantage when they pinch out terminal buds to make a plant busier or to get more flowers on something like a fall chrysanthemum.

Some perennials have lateral buds that remain dormant and they form single-stemmed taller plants. Other perennials allow most of their lateral buds to grow into side stems, forming bushy plants. This is mostly controlled by genetics and selective pinching by gardeners.

Tomatoes form a lateral bud at each leaf node and all of these will grow to form new side stems, which in turn form lateral buds and even more side stems. This results in a big unruly bush. Some gardeners prefer to remove these side branches, a process called suckering. There is an ongoing debate about this process. Should you or should you not sucker tomatoes? For a complete discussion of this topic see this YouTube video: https://youtu.be/1p6TC4-hj5E

Overwintering

Stems take two approaches to survive a cold winter. They either die back to the ground or they become woody. I'll discuss woody stems in a separate chapter.

When the temperature gets too low, the stem of a herbaceous plant dies back. During this process the plant moves nutrients, photosynthates and water back into the root system. The leaves and stems slowly go yellow, then brown and finally become less turgid. On many plants they fall to the ground where they protect the crown of the plant in winter and this is one reason why gardens should not be cleaned up in fall. In other cases, like ornamental grasses, they remain upright, providing a good winter display.

While all this is happening, the crown of the plant is developing new buds underground. These will remain dormant during winter and will become the new terminal buds in spring.

Lack of Dormant Buds

If a herbaceous plant loses the top growth and it does not have dormant buds underground one of two things happen. In some cases new dormant buds will develop and start to grow. In other cases, the plant dies.

Many plants can form new dormant shoot buds on roots. Gardeners take advantage of this when they create root cuttings. A short piece of root is planted and after a while it forms a new dormant bud which develops into a shoot.

Dahlias can be tricky to divide. Late in the year they form dormant buds (eyes) underground, right at the base of the stem. You have to split last year's stem so that at least some of it remains on each tuber to ensure that each piece has at least one eye. A tuber without a bud will die.

In some cases the dormant buds underground die in winter, either because of cold or because conditions are too wet. Without buds, the plant usually dies.

Light Affects Stem Growth

The amount and type of light also affects stem growth. You will know this if you have ever started seeds indoors using less than ideal light. Such seedlings have thin weak stems—what we call leggy plants. In the hope of reaching more light the stem cells elongate more instead of adding width. These thin stems will fail to produce strong plants.

If you start those same seeds under stronger lights or even outdoors in a process like winter sowing, you will get short seedlings with strong, stiff stems.

You will see this same effect if you place sun-loving plants in shady conditions. They grow taller, their internodes become longer and the stems never develop their true thickness. They are not getting enough light to develop properly.

Stems are also able to sense the amount of light. If one side of the stem is shady and the other is sunny, the stem will grow towards

Trees leaning towards the light.

the light. This is evident in this picture above showing the leaning aspens. The right side of the picture faces south and gets a lot of direct sun while the left side faces north and is shaded by other trees. The trees are clearly growing towards the light.

Plants, especially woody plants will always grow crooked and should be moved as soon as you see the problem.

I had the same problem in a perennial bed I built. The pathway was on the north side of a sunny bed. All of the plants grew well there, but they tended to grow towards the sun and, in particular, all flower stems faced the sun. A visitor walking along the path only saw the back of the flowers. To solve this I removed the pathway and replaced it with larger shrubs.

The Importance of Photosynthesis

For most plants, leaves are the main point of photosynthesis but stems also play an important role. They contribute to both the growth and size of the plant.

Some plants, like cactus, which are mostly expanded stems, rely exclusively on the stem for photosynthesis. Other plants, like daylilies and grasses, have almost no stems and use leaves instead.

Plants that routinely lose their foliage to herbivores such as deer or caterpillars, or to men wielding lawnmowers, rely heavily on stems to photosynthesis until new leaves can grow.

Stems also store the sugars created by photosynthesis, putting less pressure on roots and allowing plants to more quickly move those sugars to new growth points in the event that the top of the plant gets damaged.

Underground Stems

Gardeners are familiar with stems growing above the ground, but they might not recognize underground stems. These function just like aboveground stems. They connect roots to leaves and flowers and using the xylem and phloem they transport water, nutrients and sugars. The only thing underground stems don't normally do is photosynthesize.

Tubers

A tuber, also called a stem tuber, is an enlarged, fleshy, modified stem used mostly for storage of starches and sugars. A potato is a well-known tuber.

They do not have a protective tunic or basal plate like bulbs and corms, but they may have a protective skin. Tubers do have dormant buds which are commonly called "eyes." These can develop into terminal stem buds or into roots.

The tuber can be cut into sections with a few buds on each one. When planted, each one of these will form a complete plant. This method is commonly used to grow potatoes.

Other common garden tubers include anemones and dahlias.

Bulbs

True bulbs are modified stems characterized by layers of fleshy scales used by the plant for storage. When they are cut in half, they reveal visible rings like an onion, which is an example of a true bulb. The dormant bud is located in the centre of these storage rings.

At the base of the bulb is a hard, flat surface called a basal plate from which the roots will grow. Lateral buds develop around the edge of the basal plate and will eventually form bublets or offsets that can become new plants when separated from the parent plant. Eventually, too many bulbs will develop from the parent plant, causing poor growth and flowering. The bulbs should then be dug up and divided.

> ### Plant Myth: Plant Bulbs after the Ground Is Frozen
>
> I see this recommendation quite a bit and it makes no sense. Besides, who wants to dig in frozen ground?
>
> Spring bulbs and corms are normally harvested in summer, dried and shipped dry for fall sales. They are able to survive dry for quite a few months, but they keep even better in the soil.
>
> Think about the above myth. Why should you wait until the ground freezes to plant them? The bulbs from last year stay in the ground. They don't need to be dug up, stored dry and replanted after a frost. They are fine being in the ground all year.
>
> Once bulbs are planted, they start growing roots in fall. The more roots they grow before frost, the larger the plant gets. Planting in late fall after a frost produces smaller plants.
>
> The one exception to planting early is tulips. They can develop a gray mold and planting early provides warmer temperatures for the mold to grow. If you have had gray mold problems before, or if it's already growing on the bulbs, plant later, but before frost so there is less chance that the mold will grow. If tulip bulbs show gray mold it is even better to discard them.
>
> Buy early and plant as soon as you get the bulbs and corms. This is one case where old plants are not a bargain.
>
> Some other advice:
>
> 1. Don't add bone meal when you plant—bulbs do not need extra phosphate and bone meal attracts rodents.
> 2. It does not matter if you plant bulbs the right way up—they know how to grow.
> 3. Deer and squirrels like tulips and crocus—grow daffodils instead.

Stems 51

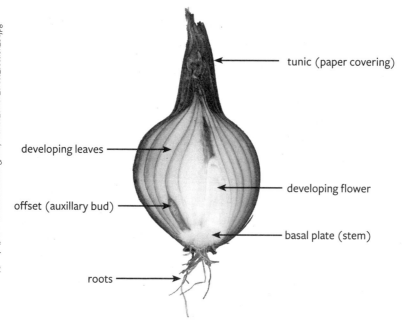

Cross section of an onion bulb.

The two main types of true bulbs are tunicate and imbricate. Tunicate bulbs have a papery covering called a "tunic" that protects the scales from drying out. Some tunicate bulbs include tulips, daffodils and hyacinths. Imbricate (also known as non-tunicate) bulbs do not have this layer. Imbricate bulbs, like lilies, should be handled more delicately and prevented from drying out.

What should you do if the tunicate falls off your bulbs as you are planting them? Nothing. The bulb will be fine.

Corms

Corms are swollen stems but unlike bulbs, they do not have fleshy scales. Instead, their insides consist of a solid mass, like a tuber.

They do have a flat basal plate from which roots develop. Corms often look like a slightly flattened or distorted sphere with a pointed tip. The pointed tip is the dormant bud and corms are best planted with the pointy side facing up. Some examples of corms in the garden are gladiolus and crocus.

Corms reproduce vegetatively by producing cormlets, also called cormels, at the base of a mature corm. Cormlets can be separated from the parent corm and planted to create new plants. They will take a couple of years to become large enough to flower.

Rhizomes

Rhizomes are often confused with fleshy roots, and the two terms are used interchangeably, but rhizomes are not roots. Rhizomes are another type of fleshy, modified stem that grows and spreads laterally beneath the soil surface. Terminal buds develop along the segmented nodes on the rhizome and grow upwards.

Rhizomes can be quite invasive since they spread underground from the original plant. If a rhizome is broken up into smaller pieces, each piece may become a new plant. Lily of the valley (*Convallaria majalis*) is a popular garden plant that uses rhizomes to spread aggressively over large areas, displacing and outcompeting less opportunistic plants. It will also invade natural areas and displace native vegetation—don't grow this plant.

Stolons (Runners)

Stolons, also called runners, are another type of modified stem that grow horizontally from lateral buds. Unlike rhizomes, stolons grow at or on top of the soil surface, but it's easy to confuse the two since

> **Plant Myth: Bearded Iris Should Be Planted with Rhizome Showing**
>
> This common myth tells you that bearded iris should be planted with the rhizome showing above the soil level or they will rot. This is not true.
>
> Several commercial iris growers in both California and Ontario bury the rhizome and they grow just fine.
>
> It is important to dig these up every three to four years and replant them so they have more space. If you don't do that, they get congested and flower less.

some rhizomes are called "stoloniferous rhizomes" if they are long and thin. Some plants that produce corms, tubers or rhizomes also develop stolons. Strawberry plants grow stolons along the soil surface that produce new shoots and roots at the stem nodes. These clones can then be separated to create new plants.

Plants with stolons can be weedy or invasive since they can reproduce vegetatively as well as seeds. Quackgrass (*Elymus repens*) is extremely difficult to remove once established since it produces many stolons that easily break off when weeding.

Specialized Stems
Some vining plants use tendrils to climb. These can be leaf tendrils as in peas and beans, or they can be stem tendrils which are used by cucumbers and grapes. In clematis the leaf petiole wraps around things to gain support and climbing hydrangea have small suckering disks at the end of a tendril.

Plants can also have twining stems, called twiners, that wrap their stems around supports as they grow. Honeysuckle and wisteria vines are twiners.

Do vines curl clockwise or counterclockwise?

Many people believe that in the northern hemisphere they curl clockwise and in the southern, anti-clockwise, an example of the "coriolis effect." This is a myth. About 90% of twining vines turn counterclockwise as you look down at the vine from above. Some, like the cucumber, curl either way, but all of this is determined by genetics and not by location.

The internet is full of claims that pole beans curl counterclockwise and runner beans twine clockwise but that is simply not true. They both curl counterclockwise.

Why should a gardener care? Many vines benefit from a gardener's help to get a vine started up its trellis and right handed people tend to start vines clockwise, which is not what most vines want.

4

Leaves

Leaves are responsible for most living organisms on earth. They convert the sun's energy to sugars, a form of chemical energy. That chemical energy drives the metabolism in every living organism on earth.

Leaves are also responsible for manufacturing a vast number of chemicals that protect the plant from insects, encourage mycorrhizal associations and attract pollinators. They even signal other plants of potential danger.

Leaves are often the most distinctive structure on a plant and many plants can be identified through the size, shape and color of the leaves more easily than by the stems or buds. Leaves also give hints to the plant's preferred growing conditions, life cycle and genetic background.

Gardeners focus on growing plants but they are really in the business of growing leaves.

Leaf Structure

The basic structure of a leaf includes the blade, a major vein that is sometimes called the midrib, lateral veins and the petiole that connects the leaf blade to the stem. The leaf margin is the very edge of the leaf and can take on many forms including smooth, wavy and serrated. The shape of the leaf and the leaf margin are key features used to identify plants.

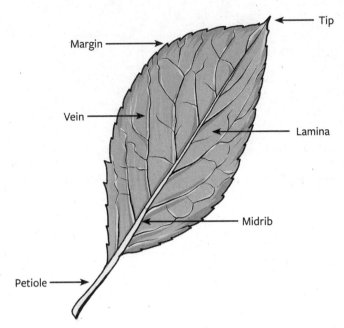

Parts of a leaf.

The tissues within the leaf are layered like a sandwich. The epidermis is the bread of the sandwich and inside the bread you find the mesophyll and veins.

The outer layer of the leaf consists of the upper and lower epidermis which holds the leaf together. Each layer is usually one cell thick, but plants in extreme climates may produce epidermises that are several cells thick. The epidermis protects the plant from water loss, extreme temperatures, UV damage, physical injuries and pathogens. To help with this protection it is covered by waxy secretions called the cuticle.

The Colorado blue spruce (*Picea pungens*) gets its distinctive color from the powdery cuticle wax on its needles, which protects the plant from strong sunlight in its native range, the Rocky Mountains.

Some leaves produce extensions of the epidermal cells called trichomes. These are similar to stem trichomes and secrete or store chemicals that deter herbivory by being sticky, smelly, toxic or bit-

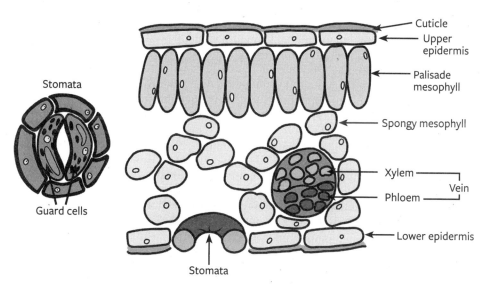

Cross section of a leaf with a close up of the stomata.

ter. They also help the plant sense insect activity and prepare a defense response. If you rub your fingers along the hairs of a tomato plant it will release a fresh, sharp fragrance. The scent is pleasant to humans but meant to deter insects and pathogens from munching on the leaves and stems.

Between the upper and lower epidermis are the mesophyll tissues which are photosynthetic. They contain chloroplasts, which produce chlorophyll, a substance that absorbs red and blue light while reflecting green light, making plants look green. There are two types of mesophyll cells: palisade mesophyll and spongy mesophyll.

Palisade cells are closer to the upper epidermis. These cells are shaped like columns and are arranged in rows like a picket fence. Most of the plant's food production takes place in the palisade mesophyll since these cells contain more chloroplasts and they are located in the upper part of the leaf.

The spongy mesophyll cells are found in the lower half of the leaf and play an important role in gas exchange, though they do photosynthesize as well. The irregularly shaped spongy mesophyll cells are loosely packed with many small spaces between. These spaces allow carbon dioxide to diffuse into the leaf and oxygen and water vapor to move out.

The veins run throughout the leaf and look very much like our blood vessels. They are large as they enter the leaf and then fan out getting smaller and smaller as they branch into many small artery-like vascular bundles. The vascular bundles are similar to other parts of the plant and contain the xylem and phloem.

Stomata

The combination of epidermal cells and a waxy cuticle makes the leaf very watertight. That is important to maintain the turgidity of the leaf and minimize the drain on roots. However, it is also important for the leaf to get CO_2 from the air and to release excess oxygen and water vapor produced by photosynthesis and this happens mostly through the stomata.

The stomata are fairly large openings on the surface of the leaf and most are located on the lower epidermis where the cuticle is thinner and there is less heating from sunlight.

Stomatal opening is formed by two special guard cells which are shaped like kidney beans and function similarly to your lips. When they are squeezed shut, no air passes through them. They can also open to let a bit of air through, or they can open fully to allow air to move easily. The opening and closing is controlled by surrounding cells which pump water in or out of the guard cells by osmosis. When fully inflated, the guard cells are closed.

Opening and closing is based on the amount of light available for photosynthesis. When photosynthesis is active the plant needs lots of CO_2 and the stomata open to provide it. When it gets dark, photosynthesis stops, which triggers the stomata to close which reduces water loss when CO_2 is not needed.

The stomata will also close when the roots sense a shortage of water in the soil. What this means is that dry soil reduces access to CO_2, which in turn reduces photosynthesis and the plant slows down growth due to a lack of food production.

Adaptation to Dry Climates

Extreme temperatures and lack of water are major barriers to plant growth. As a result, plants have adapted different leaf structures to cope with these challenging conditions.

Cacti and succulents come to mind when thinking of plants in dry climates. These plants are so successful in deserts because of numerous adaptations. The leaves and stems contain water-storing tissues that give the plant a "juicy" appearance. Relatively large quantities of water are stored in the leaves, stems and roots that the plant can access during extended periods of drought. Many water-storing plants will also produce a very thick, waxy cuticle to further conserve water.

Cacti use stems to store water and photosynthesize. Their leaves have been reduced to spines that protect the plant from herbivores and provide shade.

Some cacti and succulents have also altered their photosynthetic cycle to limit water loss. If the stomata open during the day in very hot temperatures too much water can evaporate from the leaves causing the plant to dessicate. Unlike other plants, these plants will open their stomata at night when temperatures are cooler to take in carbon dioxide and release oxygen. The plant will then store the excess carbon dioxide by converting it into a substance called malate. During the day, the stored malate will be converted into food using sunlight. Botanists refer to plants that have this unusual strategy as CAM plants, standing for Crassulacean Acid Metabolism, after the crassula plants which were used to study this altered photosynthetic cycle.

Needle-leaf plants like spruces, firs and pines are another example of adaptation to dry climates. Though there may be plenty of precipitation in northern climates, a significant portion consists of snow which is largely inaccessible to the plant until it melts. Coniferous plants adapted by producing small, thin leaves with less surface area to limit water loss. These leaves also tend to produce a very thick, waxy cuticle. They maintain their leaves year-round because their environments are not warm, moist or sunny enough to support the production of new leaves each year. The downside to less surface area on needle leaves is that the leaves can't photosynthesize as effectively as broad leaves, resulting in slower growth compared to deciduous plants.

Cuticle Damage
The cuticle plays a major role in preserving water and protecting the leaf from pathogens and insects. It is critical to plant health. It can also be damaged by the actions of unsuspecting gardeners.

Gardeners like to use home remedies and many of these concoctions include soaps and detergents of various kinds. Soaps and detergents are designed to remove oily and waxy compounds. Spraying them on plants can remove all or part of the cuticle. This makes the plant vulnerable to infection and more insect damage.

A drop or two in a bottle of mixture acts like a surfactant and helps it stick more efficiently to plants. It is unlikely to cause much of a problem. But some mixtures use more soap and then it can cause harm.

The reason soap is used in these mixtures is that gardeners have learned that insecticidal soap is a good way to control insect pests and they assume household soap will also work. It might, but it also harms the plant you are trying to protect.

Insecticidal soap is made from potassium salts. Home soaps (most bar soap and hand liquid soap) are mostly made from sodium salts which are much more harmful. Most so-called dish soaps, including the ever popular blue Dawn dish soap, are not even soaps. They are detergents and are even more harmful to plants than soaps.

How Sun Affects Leaves

Leaf growth is affected by the amount of light the leaf gets. Plants growing in full sun tend to form leaves that are smaller, thicker and have more protective waxy coverings or hairs. Plants in shade have more difficulty getting light and therefore grow leaves that are more efficient at trapping light. They are larger, thinner and have less surface protection.

These differences can even be seen on single plants. The inner leaves on shrubs tend to be larger, while the outer ones are smaller.

In low light conditions the plant will focus its growth on height and producing more leaves, while sacrificing flowering and fruiting. It will elongate stems to grow as tall as possible so leaves are closer to the light source.

The European buckthorn has become very invasive in North America in part because of its strategy for using sunlight. It leafs out early in spring before taller trees steal all the light, and it still has green leaves well after most native deciduous trees have dropped theirs. In effect, it has a longer growing season than natives and the extra time under strong light allows it to grow in very shady conditions.

These light effects can also be important when moving plants. If you move a plant growing in full sun to a shadier spot, its existing leaves are conditioned to more light and they will struggle. On the other hand if you move a plant from shade to full sun, there is a good chance the leaves will burn because they have not developed a thicker cuticle and light-trapping pigments for the high light conditions.

Trees and shrubs that have been sitting at the nursery with one side in sun and the other in shade should ideally be planted in the same orientation in your garden.

In each of the above cases, the new leaves that form on the plant after it is moved will adapt to their new location and be able to handle the change in light. This is another good reason to move deciduous plants in early spring or fall when leaves are not yet formed.

Consider what happens when a large tree dies or is removed in a shade garden. Suddenly, the shade plants underneath the tree go from shade to full sun. A lot of the leaves will burn. The shift may be dramatic enough to even kill some plants because they can't adapt fast enough.

You can help with sudden increases in light by providing artificial shade and by keeping plants well watered. If you know you have to remove a tree, consider doing it in late winter, before new plant growth. That will make it much easier for plants to adapt. I did precisely that when I removed a large sugar maple from my shade garden. I removed it in winter and kept things well watered the following year. Not only did the plants survive, they actually grew better than in full shade.

Why Are Some Leaves Red?

Most plant leaves are green, some are red or reddish-purple, and even green leaves can take on a red tinge in certain conditions. What causes this red coloration?

The red color is caused by a group of red-colored chemicals called anthocyanins. They are present in all leaves and serve various purposes. They act like sunscreen and protect the chloroplasts from too much light.

Leaves that have low levels of anthocyanins still look green because the amount of chlorophyll is higher. As the amount of anthocyanin increases, the red color intensifies and the leaf starts looking redder.

Some plants naturally have red leaves, due to high anthocyanin levels, but that causes a problem for them. Firstly, making these chemicals is energy intensive for the plant. Secondly, they reflect red light and absorb blue and green light, reducing the amount of light available for photosynthesis. The net effect is that red-leaved plants grow slower. A red-leaved canna lily grows slower than a green-leaved one.

Red-leafed cultivars tend to be greener in shade. The shade has less light making it more difficult to grow. These plants react to this by reducing the amount of anthocyanins they produce and therefore look greener. This effect is dramatic in the concorde barberry (*Berberis thunbergii* 'Concorde') which is covered in deep purple leaves, but if you lift up the upper leaves you find that all of the shaded leaves underneath are green.

> **Plant Myth: Evergreen Needles and Oak Leaves Are Acidic**
>
> The claim that pine needles are acidic is very common and many people gather them up and add them to their garden to make the soil less alkaline. They are claimed to be perfect for rhododendrons and blueberries which need acidic soil to grow.
>
> Some online claims include a reference to a professor who supposedly studied this and found them to be acidic. I didn't believe it, so I contacted her. She told me that she gets at least one request a month for the study, but that she had never studied the pH of pine needles. Someone made up this rumor and all the other bloggers just repeat the nonsense.
>
> Evergreen needles are slightly acidic when green and on the tree. Once they fall off, they are not acidic. The same goes for oak leaves.
>
> Even if they were acidic, they still would not acidify your soil because of its buffering capacity. If you want to understand soil and its pH better, have a look at my other book, *Soil Science for Gardeners*.

The development of red coloration in normal green leaves can also indicate an abiotic (environmental) stress. Low levels of phosphorus will cause the plant to move phosphorus from older leaves to new ones, which results in older leaves becoming red. Dry conditions make it more difficult for plants to move nutrients and sugars around and can result in redder leaves. A drop in temperature has a similar effect.

Fall colors are due to less chlorophyll production as the leaves shut down and an increased production of anthocyanins to deal with the colder temperatures. Other pigments like orange and yellow carotenoids are also involved.

Functionality of Damaged Leaves

A rose develops some black spots on the leaves or they are partially chewed off by pests. Should the leaf be removed?

A lot of people remove damaged leaves from plants as soon as they see them and in most cases, this harms the plant.

Green leaves are used by plants to produce food. Even half a leaf is better than no leaf. Each time you remove a green leaf, it reduces the ability of the plant to produce food for future growth.

What about diseased leaves? Many fungal and bacterial diseases spread across the leaf once they start. A black spot on a rose gets larger and soon there might be two or three spots. Eventually the whole leaf might be covered. You've seen the same happen with powdery mildew. A small speck of white will eventually cover the whole leaf.

A leaf that is partially damaged by disease will still be productive to the plant. Once it is mostly covered with disease, it can be removed with little loss to the plant.

There is also the concern that an infected leaf will infect other leaves. By the time you see these diseases most of the leaves are infected or at least have spores on them. The spores are also in the soil, on the stems and even nearby plants. Removing a leaf is unlikely to slow down the progress of a disease.

Natural Pesticides

Plant leaves contain sugars, protein, water and all kinds of minerals, making them a perfect food for all kinds of things from insects to rodents to humans. To combat this problem plants produce many chemicals to make their leaves less palatable. You are probably familiar with the milkweed that contains a toxic latex-like compound that keeps almost everything from eating it. The exception is the monarch butterfly which has developed the ability to tolerate it.

All plants produce a wide range of similar chemicals, collectively called natural pesticides. There are thousands of these chemicals in every plant and each time we eat any part of a plant, we ingest these pesticides. On average we eat between 5,000 and 10,000 different natural pesticides, most of which have not been adequately studied by science, so we don't know how dangerous they are. What we do

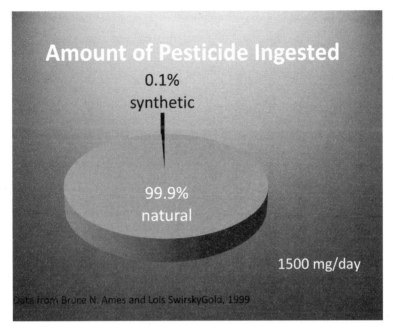

Amount of natural and synthetic pesticides ingested each day.

know is that humans have been eating them for generations without significant problems. We don't need to get alarmed about these numbers.

A lot of people are concerned about eating synthetic pesticides and it's one reason organic food has become so popular.

When we compare the amount of synthetic pesticides we ingest to the amount of natural pesticides, we find that the synthetic ones make up less than 0.1%. Society has been falsely conditioned to worry about this minor component of the pesticides we consume. Don't get me wrong, we need to be careful with synthetic pesticides and use them as little as possible, but they are not as significant in our diet as you've been led to believe.

You might be thinking that natural is better and safer but this is another common myth. Many of these natural pesticides are carcinogens and some of the most toxic chemicals on the planet are natural compounds produced by plants. Take ricin for example. It is found in the beans and plant material of the castor bean, a common garden plant, and it is one of the most toxic chemicals on earth.

Signaling Between Plants

Imagine a caterpillar chewing on a leaf. The plant seems completely helpless, but that is far from being true since it uses chemical warfare to protect itself.

The chewing action of the caterpillar causes the damaged leaf to produce more natural pesticides, some of which will make the leaf taste bad. It then produces and releases a set of volatile signaling compounds. These chemicals float through the air and land on other leaves, both the plant's own leaves as well as the leaves of surrounding plants.

When these signaling compounds land on a leaf, they trigger the production of more pesticides. Very quickly, all of the leaves in the area are more toxic to the attacker.

In effect, the plant has sent out a signal that says, the bugs are coming. The war between plant and bug has begun.

These signaling compounds work best on other plants of the same species and they may also work on some of the other species.

Some people, including scientists, describe this process as "talking plants," or "communication between plants." I think that is incorrect. Communication implies the transfer of information from one individual to another such that the receiving party understands the message and that a message was sent. This all implies thought and knowledge, neither of which is present in plants.

Don't get me wrong—what plants do is quite spectacular, but there is no thought or knowledge involved. A plant will make the signaling compounds even if it is the only plant in the room. The response is automatic and is a result of simple chemical reactions. The chewing action causes a chemical change in the leaf, which results in the production of signaling compounds. When they land on another leaf, they cause a new set of chemical reactions. This description of events is not as glamorous as "communicating" plants, but it is still a phenomenal process.

Pest-Proof Leaves

Plants use a variety of strategies for protecting their leaves from pests and a smart gardener picks those with pest-resistant leaves.

If you grow a number of hostas you soon realize that some are slug magnets and others are never touched. You can easily see this in fall by looking at the condition of the leaves. Some hostas have thinner leaves making them tastier. Others are thick and waxy. The ones with big blue leaves seem to be quite pest free.

Plants with fuzzy hairs on them are also unpalatable to most pests and even deer leave them alone until they are really hungry. Some leaves produce foul-tasting chemicals like the common milkweed's.

These protective measures apply to all plants. Select the right ones and you will have fewer pests.

Water Stress and Wilting Leaves

The sun not only provides higher light levels but it also warms the leaves. Both of these will increase the rate at which leaves lose water through transpiration, putting a strain on roots to pick up enough water.

As leaves lose water, the water pressure inside them drops. The plant manages this by releasing water from special water-holding organs called vacuoles. These get smaller as they release water to cells for other more important purposes. At this stage there is no permanent damage and the plant can easily recover.

As water levels continue to drop, permanent damage starts, first in leaves and flowers, then in stems. Severely stressed plants then abort damaged parts; flowers fall, leaves dry up and fall off. Eventually even stems collapse. The last to go is the root system and crown of the plant.

Most leaves are adapted to some water shortage and show no signs of distress. Others, especially large-leafed plants, can show the extra stress by wilting.

Many take this as a sign that the soil is dry and needs to be watered. That is certainly a possible explanation, but with some plants, like ligularia, this happens every afternoon no matter how wet the soil is. These plants have large thin leaves and can't move enough water into them on a hot day to keep them turgid. As the sun sets, the leaves cool, stomata start to close, water loss is reduced and the leaves become turgid again. For these plants it is a normal daily cycle.

I've tried growing ligularia in full sun with their roots sitting in a wet bog and in part shade. In both places they get droopy leaves in summer when it's warm. Watering makes no difference.

Wilting of leaves also happens when a plant is moved in summer. Roots are reduced in size so they can no longer provide enough water for leaves, which causes wilting. This happens even when soil is completely wet.

Wilting can occur even when a plant is not moved due to dry soil, dry air or high temperatures. Watering only solves the dry soil condition.

Many gardeners respond to these problems by watering more or watering more frequently. But that can reduce the air in the soil leading to root rot, which then makes the problem worse.

Take the time to understand why the plant is wilting and water correctly. If the plant was moved, provide some shade to reduce water loss and wait for new roots to grow. If the plant was not moved and temperatures are hot or the air is very dry, there is not much you can do.

In all cases water correctly. When you water, water very thoroughly and wet soil even a foot or more away from the plant. Mulch to maintain moisture. Then don't water again until the soil starts to dry. You can tell when the time is right by sticking your finger in the soil.

Leaf Abscission

We tend to think of leaves as being a permanent fixture on plants but they aren't. A leaf is a temporary organ that has a specific life span. Once it is no longer required by the plant, or when the plant can no longer support it, it's dropped. This process is called leaf abscission.

The leaf is connected to the stem at the petiole. The xylem and phloem from the stem connect directly to the xylem and phloem in the leaf. During abscission, these vessels need to be severed along with other connecting tissue. It is not unlike having a limb removed surgically, except that the plant performs this on its own.

As a leaf begins to grow, a special set of undifferentiated cells develop, called the abscission zone (AZ). The AZ is usually formed at the start of the petiole right next to the stem. Normally, the auxin hormone levels are kept high to keep the AZ from developing further. When it is time to lose the leaf the plant reduces the auxin levels which cause changes in the AZ.

Nutrients stop flowing into the leaf and some are removed so they can be used in other parts of the plant. The cells in the AZ differentiate into two distinct layers, one on the leaf side and one on the stem side. The stem side accumulates a chemical called suberin, which is a key component of cork. This helps plug the open ends of the xylem and phloem, eventually developing into a solid structure called the leaf scar. By corking the xylem and phloem, the plant prevents pathogens from entering the vascular system and infecting the plant.

If you look at older leaf scars you may still be able to see the remnants of the vascular strands.

It is always better for you to let the plant drop its own leaves. If you cut them off or pull them off, the plant can't immediately close off the wound. This allows pathogens from the air, water or dirty tools to enter. There are a wide variety of plant diseases called vascular wilt or vascular dieback, which involve pathogens entering through the xylem. These bacteria and fungi proliferate in the xylem and block off the vessels throughout the plant, similar to when human arteries get clogged with fat. When this happens the rest of the plant is unable to get water and nutrients.

There are a number of reasons why leaf abscission happens.

Limited Lifespan

Leaves on a plant have a limited lifespan and this aging process is called senescence. Many plants naturally lose older leaves. For example, the palm has a set of new leaves at the top and a long trunk-like structure that clearly shows the leaf scars where older leaves had been attached.

Many plants drop the lower leaves as they get taller.

Fall Leaf Drop

Fall leaf drop in deciduous trees in temperate climates is a well-known example of leaf senescence and abscission. Changes in night length and temperature trigger leaves to start the process. The green chlorophyll degrades, revealing other existing pigments underneath (carotenoids and anthocyanins). Nutrients and sugars are removed from the leaf and abscission begins.

Abiotic Stresses

A number of abiotic stresses can also cause leaf drop. In drought conditions a plant will drop its leaves when it can no longer support their water and nutrient requirements. There is no point in keeping an appendage that is not productive.

When plants are newly planted they can be very stressed due to root damage. This is no different than a drought condition. The plant may react by losing some or all of its leaves. It is not uncommon for trees to drop all their leaves and then regrow them a month later.

Dropping Lower Leaves

Lower leaves tend to be less effective at photosynthesizing since they become shaded out by the plant's upper leaves. Like humans, plants are always looking to maximize their gains, and it's more efficient for a plant to drop its lower leaves and allow nutrients to move into the more productive upper leaves. In essence, the plant is simply pruning itself. The process is slow and steady, with the plant dropping one leaf at a time.

If the plant is losing leaves rapidly, dropping multiple leaves at a time or dropping newer leaves, the leaf loss is probably not natural. The plant might not have the right conditions to support new growth, or may be damaged, diseased or ravaged by pests. Some nutrient deficiencies, like nitrogen deficiency, start at the lower leaves before moving up to the rest of the plant, so it's good to keep an eye out on any sudden leaf changes.

Variegated Leaves

Plants with variegated leaves are highly prized in the gardening world because they are more visually interesting than plants that are just green. They have leaves with more than one color and can be two-, three- or even four-toned. Colors can even change throughout the seasons.

Variegation is a function of the amount of chlorophyll and other pigments in any given cell. White patches are cells with low chlorophyll and reddish ones have low chlorophyll and more red pigments. The hosta is the classic plant that has been exploited to produce a wide range of variation types, but many other plants also show this trait.

Where do these plants come from? It is usually a mutation. Seedlings that have no chlorophyll are relatively common, but finding one with green sections and colored sections is rare. If a seedling is all white, it will certainly die. With no chlorophyll the plant has no way to make food.

When shopping for plants be aware of the fact that variegation might be seasonal. A great variegated plant in spring might be quite green by summer. This is the case, for example, with the variegated Japanese lilacs 'Chantilly Lace' and 'Golden Eclipse.' Before you purchase a variegated plant, check to see if the color lasts all summer.

Such plants will be variegated again next spring. Apparently, the change is due to warmer temperatures in summer.

I have an astrantia seedling that starts out in spring with no green on the leaf, but by midsummer it is fully green. If it stayed white all summer it would probably not survive. On the other hand

Caryopteris 'Snow Fairy' emerges in spring with nice partly white variegation, and it is still strongly variegated by late fall. This is a great perennial.

Inherited Variegation

Variegation is rarely an inherited trait that is passed to offspring grown from seed. In most cases the offspring will be green so these plants have to be propagated vegetatively.

Leaf variegation is usually detrimental to the plant since chlorophyll is needed for photosynthesis and such plants are often less vigorous than their green cousins.

Reversion

Variegation can be quite stable or very unstable. If it is unstable, some new growth will revert back to green, a process called reversion. If this green portion is not removed it will grow more aggressively than the variegated part and it will become the dominant growth. Eventually you will lose the variegation.

The Harlequin maple (*Acer platanoides* 'Drummondii') is a great-looking tree, but almost every one I have seen has partially reverted back to a solid green. This reversion needs to be removed which might require an annual visit from an arborist. Oddly enough, this mutation of the Norway maple seems little prone to tar spot disease, a common problem with other Norway maples in many areas.

Hostas can also revert over time. If this happens you need to divide the clump and remove any section that is reverting. If in doubt, remove more than you think you need to remove.

Effects of Light and Nutrients

The amount of variegation and the intensity of the non-green colors can be affected by light levels. In low light levels such plants have trouble making enough food so they increase the number of chloroplasts which makes the leaf look greener. However, this is not always the case. For example, in variegated peperomia, a common houseplant, less light results in more variegation.

These effects due to light are not reversion since they are temporary and can be reversed with a change in light intensity. Reversion is a genetic change that can't be reversed by cultural changes.

Gardeners who experience a change in color on variegated plants frequently blame nutrient levels but it is unlikely that such changes are the result of a nutrient deficiency.

Variegated plants can be tricky—they burn when exposed to too much sun yet are intolerant of too much shade. They might grow more slowly or have smaller leaves. But variegated plants will always be valued by gardeners for how well they pop out in a sea of green leaves.

5

Flowers

Flowers are an important goal for both plants and gardeners. They create beauty, produce fruit and provide a way for the plant to reproduce. They are also a critical part of the ecosystem and provide vital food for pollinators.

Parts of a Flower

All the hard work that goes into researching, planting, watching and maintaining gardens is well worth it when flowering plants bloom in a symphony of color. Flowers generally consist of four parts: the petals, sepals, pistil and stamens. The petals and sepals are the non-sexual parts of the flower and have functions unrelated to reproduction. The pistil and stamen are the sexual parts that play starring roles in reproduction. Flowers with all four parts are called "perfect" flowers, but there are many "imperfect" flowers that lack one or more of these parts.

The sepals are usually green, leaf-like structures that enclose and protect the flower bud before it blooms. They can also provide support for the flower after it opens and they may have hairs or thorns that provide extra protection. Sepals are considered modified leaves and can photosynthesize while they cover the bud. Once the flower opens, sepals may remain on the flower or they may wither and fall off.

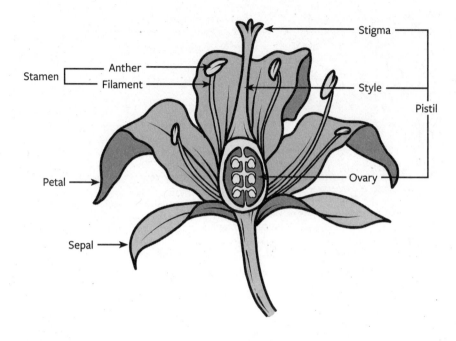

Parts of a flower.

Petals are normally the showiest part of the flower and what gardeners tend to value most. Technically, they are modified leaves that surround the reproductive parts of the flower.

Some flowers have sepals that change color and shape to look more like petals. The true lily has sepals that look exactly like petals. In the German bearded iris the sepals become the "falls" of the flower and can be showier than the petals. When sepals look more like petals, both the sepals and petals are collectively called tepals. Many flowers with tepals are highly desirable garden plants because the flower is showier.

Petals are really only for show, since the most important parts of a flower are the reproductive parts. These parts can either be male or female, and a plant can have either or both.

The male part is referred to collectively as the stamen, which includes the pollen-producing anther that is supported by a slender stalk, called a filament. Filaments can be long, short or not present

German bearded iris with vertical pink petals and red falls (sepals).

at all since the anthers are the key part for reproduction. Filaments provide nutrients to the pollen as it develops and they position the anther where the pollinators are most likely to contact the pollen.

The female parts are collectively called the pistil, and include the stigma which receives pollen, the ovary and a tube that connects the stigma to the ovary called a style. The ovary contains one or more ovules each of which contains one egg cell.

Pollination

Angiosperms are plants that produce flowers, fruits and seeds as a way to reproduce. This group includes 80% of the plants on earth and includes most garden plants as well the vast majority of all food plants we eat, including grains, fruits and vegetables. Pollination is critical for both their survival and ours.

The simplest pathway for pollination is as follows:
- Pollen grains from anthers land on the stigma. If the stigma is mature it's sticky, and the pollen grains will stay on the stigma.

- A pollen grain contains one tube cell and one sperm-producing cell. The tube cell will produce a tube that grows down into the style and eventually reaches an ovule in the ovary.
- The sperm cell then travels down the tube and reaches an egg in the ovule. The sperm fertilizes one egg, which develops into a seed.
- After fertilization, the ovary develops into a fruit which protects the developing seeds.

Of course, there are some nuances in the process. Both the pollen and stigma go through a maturation process and are only functional when mature. Immature or old pollen will not work, even though pollen can last many years in the right conditions. The stigma needs to be ready to accept the pollen and it may only be receptive for a few hours or days.

The pollen and stigma also need to be compatible. The pollen tube and style exchange genetic information, and if the plant believes the pollen is incompatible (from a different species, a very close relative or an unfit partner), the stigma will release a toxin that prevents the pollen tube from growing. In nature, plants generally prevent inbreeding or hybridization between species so they can produce the fittest offspring and maintain the species.

What Causes Flowering?

Plants spend most of their effort to grow larger and only when certain criteria are met do they flower. It is important for a gardener to understand these criteria.

Plants need to be large enough to flower. In annuals this happens very quickly, but some trees might need to be 10 or even 20 years old. Many large oaks do not bloom until they are at least 20 years old but my dwarf chinquapin oak has acorns on it after only five years from seed.

Plants also need to be healthy and have access to enough light. Anything that prevents a plant from growing to its full potential may prevent flowering.

Even when all of these criteria are met, the plant still needs certain triggers to flower. One of the most common is day length. We commonly talk about day length, but plants actually sense the duration of darkness, not light. So it is more correct to talk about dark length, rather than day length.

Plants fall into one of three categories: long night, short night or night neutral.

Chrysanthemums are long night plants that need a dark period of 12 hours or more. This is the main reason they don't flower until fall when days are shorter and nights are longer. These plants can be tricked into flowering any time of the year by providing a long night.

Roses are night neutral. They start to grow in spring and once they are large enough, they flower. Some roses are repeat bloomers and they continue to flower until frost. Others only bloom for a few weeks. The duration of the bloom is determined by genetics.

Fruit trees bloom in spring when there is about 12 hours of darkness. In temperate zones, this amount of darkness also happens in fall, but the trees don't bloom then because there is a second trigger involved, called vernalization. These plants need two things to form flower buds: a 12-hour dark period as well as a previous cold period.

In order for plants to sense the duration of darkness they have to be able to measure time. This is a complex process that uses molecules called phytochromes. Far-red light changes the phytochromes, which in turn triggers bud initiation.

Tropical plants are not exposed to changes in day length, nor to cold periods and must use a different system to control flowering. Science is still working out the specifics, but it seems some tropical plants use the natural cycles of solar radiation. At the equator, radiation intensity peaks when the sun passes directly overhead at the spring and autumn equinoxes, and wanes towards the solstices. For example rubber trees tend to bloom at the same time, and have two blooming periods, one in spring and one in fall.

Rainfall can be a significant trigger for some plants. Desert plants tend to bloom in spring, but they wait until right after a significant rain event.

A Flash of Light at Night

In nature this process of monitoring darkness works quite well provided the plant gets continuous darkness. Light from the stars or even a full moon is too weak to affect it, but this is not true in our urbanized world. Light from streetlights and home lighting can interfere and prevent the plant from getting enough darkness.

Christmas cacti are long night plants and normal home lighting is enough to interrupt the dark period, which in turn prevents flowering. Street lights can interfere with flowering of ornamental plants such as pineapple sage. Control of light during the dark period is critical for the flowering of some houseplants.

Many people are now also adding garden lighting and outdoor home lighting and these can also interfere with blooming if the light level is high enough and if it is the right wavelength of light. The table below shows how different types of light affect plants. Note that it is the red light that is most critical.

Most garden LED lights have a low light output that won't affect plants but if you use these try to get ones with low red light.

Lights that are on for only a few hours after dusk are better than ones that stay on all night.

Light Source	Wavelength	Potential Effect
Florescent	High blue/low red	Low
Incandescent	High red	High
Mercury valor	Violet/blue	Low
Metal halide	Green/yellow/orange	Low
High pressure sodium	High red	High

Source: William R. Chaney, Department of Forestry and Natural Resources, Purdue University, West Lafayette, IN 47907 https://www.extension.purdue.edu/extmedia/fnr/fnr-faq-17.pdf

Why Do Plants Not Flower?

Most plants are grown for their flowers and it is extremely frustrating when they don't flower. There can be many causes and it can be difficult for a gardener to decipher the problem. The following are some of the more common reasons why plants don't flower.

Plants Are Not Mature

Plants prioritize the use of their energy and reproduction is low on the list. Perennials usually don't flower for several years. Crab apples take seven years and oaks take several decades. The century plant (agave) needs to reach a certain size before it will bloom. If a plant is not mature, it uses the energy to grow and there is not much you can do to speed up the process.

Incomplete Vernalization

Plants that require vernalization need to have a cold period that is long enough as well as a temperature that is low enough and this varies from plant to plant. Some plants might grow just fine in warmer climates, but might not flower after a warm winter.

Many spring bulbs need a period of vernalization. In warm climates this can be accomplished in a refrigerator. Give them a cold treatment of 35°F (2°C) to 48°F (9°C) for a minimum of 12 weeks. They can then be planted for a spring bloom. You will need to dig them up each fall and repeat the process.

Plant Myth: High Phosphate Grows More Blooms

Many gardening experts promote the idea that a high phosphorus level will promote more flowers, but this is a myth. Plants need all of the nutrients to produce flowers and fruits. In particular, there are minimum levels of potassium and nitrogen that are needed for fruit set because they contain higher levels of potassium than other plant parts. Lower levels of it, and even low levels of boron, can prevent fruit set.

All parts of the plant need all the nutrients. An excess of one nutrient, like phosphate, does not make the plant grow better, nor does it cause a plant to bloom more. Don't waste your money on bloom booster type products.

Some spring bulbs can be purchased "pre-chilled," saving you the trouble of this step. However, if a pre-chilled bulb is left out in warm temperatures for long enough, it will de-vernalize and not bloom.

Winter Kill of Buds

Trees and shrubs that bloom early in the year such as forsythia, lilac, magnolia and some hydrangea tend to form their buds in late summer. A particularly cold or wet winter can kill the buds.

Even more deadly to buds is a late winter warm spell followed by very cold temperatures. The warm period wakes up the buds and they start to grow. Once this happens they are more easily damaged by cold.

Some plant groups are very confusing to gardeners and this includes the hydrangea. There are several different types of hydrangea; some form buds in fall on old wood, and some in early summer on new wood. The bigleaf hydrangea (*Hydrangea macrophylla*) is very popular and it forms buds in late summer so it can flower the following spring. It is sold as a zone 5 because the roots and stems are hardy in zone 5. Unfortunately the buds are not. This means that in zone 5, and even zone 6, buds are killed in winter leaving gardeners wondering why their plants don't flower.

Check out this link to identify the type of hydrangea you have: https://www.gardenfundamentals.com/hydrangea-identification/.

Some magnolias bloom very early in spring and others later in spring. The former group fails to flower properly in very cold years, while the latter group flowers just fine. Picking the right cultivar can have a big impact on flowering success.

Control the Fertilizer

Plants receiving too much fertilizer, especially nitrogen, will produce a lot of lush growth, weak stems and fewer flowers. Too much phosphate can tie up other nutrients like iron and also reduce flowering.

On the other hand, a low level of nutrients results in poor growth and few or no flowers.

If you use synthetic fertilizer, get a soil test so you know what to apply. Alternatively, just top dress with an inch of compost each year and that will provide enough nutrients for most soils.

Poor Growing Conditions

Plants that are growing well flower better. Anything that limits plant growth will also affect flowering and this includes drought, shade and high or low temperatures.

Having said that, some stress might help plants to flower. If an orchid does not flower I recommend that you stress it a bit. Skip a fertilizer cycle, keep it drier or give it cooler night temperatures. A cymbidium orchid that does not get a touch of frost won't flower well. Some of these stressors might work, but only for plants that are growing well and are healthy.

I have been growing orchids for 30+ years and in my experience a bit of stress might work, but a large well-grown orchid that is receiving the right amount of light blooms every year without a stressor.

Mast Years

Many plants will have a year in which they produce a lot of flowers and follow this with a lean year with few flowers. This is very common with trees.

The various types of fruits and nuts produced by trees and shrubs are collectively called "mast" and when a particular species produces a lot of mast, it's called a mast year. These mast years are important to ensure offspring since the amount of food produced in these years is too much for the hungry rodent and bird populations, allowing a good number to germinate.

This does mean that flower production will be better in some years and this can lead to the development of new gardening myths. If you tried some new concoction—maybe weed tea—and you got a lot of flowers you might incorrectly attribute that to the weed tea. That is why such research always uses controls and is usually carried over for several years.

We had a crazy spring in 2020. It got warm quite early and then we had a very cold spell. I noticed all the spring shrubs bloomed especially well. Magnolias were covered with flowers. The forsythia were the best in ten years. You might be thinking that the weird spring weather had something to do with this, but remember, most of these buds were formed the previous summer, and their growth is very dependent on plant growth in the year before that. As gardeners, we just don't have the data to explain these events. Just enjoy them and expect fewer flowers next year.

Incorrect Pruning

I mentioned above that some woody plants produce buds in fall and some in late spring. It is important to know what your shrubs do before you prune. The best time for pruning is late winter but if you prune spring-flowering shrubs at this time you will remove buds. For this reason, most people prune spring-flowering shrubs right after flowering.

In the case of lilacs, the new buds start forming a few weeks after flowering is done so you have a fairly small window in which to do your pruning.

Tough Love for Plants

Flowering and seed production require a significant amount of energy and there are times when a gardener should prevent flowering.

Many plants in flower are sold with an undersized root system growing in cramped small pots. They are often heavily fertilized to maximize growth and blooms. You now take this plant home and do some root damage as you plant it. The poor plant thinks it will die and tries its best to flower and make seeds.

The best thing you can do for this plant is remove all buds and flowers as soon as you get it. This prevents the plant from wasting energy on flowers and allows it to reallocate its energy to root growth. You sacrifice this year's blooms for a healthier plant in the future.

I do this with most plants. If I really want to see the bloom, I allow one or two flowers to open and remove the rest. Once I have seen the flower, I remove it so it does not get pollinated.

Attracting Pollinators

From the plant's perspective, flowers only have one goal: to attract pollinators. They do this very well in a number of ways and each flower has its own bag of tricks.

Some flowers produce a lot of nectar, a sugar-rich liquid that is a favorite food of bees, butterflies, hummingbirds and many other insects. Others provide a lot of nutritious pollen which is a great protein source for many pollinators. And still others don't bother with either of these and rely solely on producing flowers which are so attractive or smell so great that pollinators can't resist paying them a visit.

Many plants adjust the scent of flowers based on light levels. For example, some orchids produce a fragrant scent only at night to attract moths. Others are fragrant only during the day, because their pollinators are not active at night.

Bees take away almost 90% of the pollen they collect and the remaining 10% ends up pollinating flowers. This might seem like a wasteful process, but it is actually quite efficient. Wind-pollinated plants produce much more pollen in the hopes of successful pollination, although it is not nearly as nutritious and therefore easier to produce.

Nectar is a favorite food of pollinators but it also attracts predatory wasps, which then lay eggs on caterpillars feeding on the plant. Everybody wins, except the caterpillars.

Color Affects Pollinators

Even with nectar and pollen, flowers still have to advertise and they do this with different colors and shapes.

Plants play favorites with pollinators. It might seem disadvantageous to rely on fewer pollinators, but plants want pollinators to be loyal to their species so that more of their pollen lands on suitable mates instead of being wasted on other species. The pollinators also have an incentive to be loyal—the more specialized the blossoms, the more likely they're still full of pollen and nectar since they haven't been emptied out by a wide variety of insects.

Flowers with blue and violet petals are especially attractive to bees, while hummingbirds prefer red, pink or fuchsia. Bright orange, pink and red flowers, like the butterfly weed, are attractive to butterflies. Night-blooming flowers like the evening primrose are pollinated by nocturnal creatures like moths and bats, and are commonly white so they can be spotted more easily in the dark.

The colors and patterns we see are only part of the story. Many insects see wavelengths that the human eye can't perceive, including the ultraviolet range. Since flowers have developed visual cues for pollinators and not us, many flowers have UV-visible markings. These cues are usually stripes or splotches that direct pollinators right to the plant's reproductive parts to facilitate the transfer of pollen grains.

Flowers that look plain and unimpressive to humans may be very attractive to pollinators.

Flower shapes have also evolved to accommodate pollinators. Deeper flowers are meant for pollinators with long tongues or beaks, like hummingbirds and some butterflies. Flowers with shorter petals or petals that fold back can be pollinated by ones with shorter tongues, like bees. If a flower has a very unusual or complex shape it's likely that the plant has a close relationship with a specific pollinator.

The striking hooded pitcher shape of jack-in-the-pulpit (*Arisaema triphyllum*) is pollinated by fungus gnats. Normally they lay their eggs on fungus, which the larvae eat. The jack-in-the-pulpit flower produces a fungal scent (dank and earthy) to fool the insect into thinking it is a good place to lay eggs. Once they enter the pulpit, they have a hard time getting out. In a male flower, they get dusted with pollen and then leave through a tiny opening at the base. Fungus gnats are not that smart so they keep trying different flowers. If they enter a female, there's no way out so they just keep crawling around, pollinating the plant until they die.

The horticultural industry often produces plants with flower colors that wouldn't exist in nature, inadvertently altering the plant's relationship with pollinators. It's important for gardeners with an interest in pollinators and sustainability to understand the relationships between flowers and pollinators.

Enjoy the Bracts

A bract is a specialized leaf that is usually located right below a flower and looks as if it is part of the flower. They can be green or any other color. In some cases they look more like petals than leaves. Here are some examples of bracts:

- The red so-called flowers on a poinsettia are actually bracts. The flower is quite tiny and located at the center of the bracts.
- Bougainvillea shows brightly colored flowers with a white center, but most of this so-called flower is the bract. The white center is the flower.

The most visible part of a poinsettia flower are the bracts.

- Hellebores are very popular because gardeners think they bloom for a long time but that is not true. The flowers are quite small and last a couple of weeks at most. The colorful bracts last for many months.
- The large circular disk on anthurium is also a bract, or more correctly called a spathe, which surrounds the fleshy spike in the center (spadix).
- The so-called flower on a flowering dogwood is also a bract.

Why should a gardener know about bracts? Simple. They can be very attractive and even more colorful than flowers and they tend to last a very long time. If you want a long blooming perennial or shrub, look for one that has attractive bracts.

Clematis Petals

I'll bet you have never seen clematis petals. They do make some of the largest and showiest flowers in the garden, but they don't have petals. The attractive parts of a clematis flower are sepals, another kind of specialized leaf.

Bracts and sepals are both leaves and botanically there is a difference, but a gardener can think of them as being the same thing.

Dioecious and Monoecious Plants

Plants with flowers that contain both male and female parts, are considered to have "perfect" flowers and are called monoecious. There are other options in nature. Some species have male and female plants and are called dioecious. In this case one plant will have all male flowers and a different plant will have all female flowers. There are also plants that have male-only flowers and female-only flowers on the same plant but these are still called monoecious.

There are advantages to both monoecious and dioecious flowers. Plants with perfect flowers simplify the pollination process because the pollen does not need to travel very far to contact the stigma. Although this process is easy it does have a downside. All the offspring get their genes from the mother plant which reduces the genetic diversity in a population. This inbreeding can lead to weaker plants long term.

Some monoecious plants overcome this problem by making the flower self-infertile, so its own pollen won't produce seeds. In other cases the flowers are created in such a way that it is unlikely its own pollen will contact the stigma. For example, one way to accomplish this is to have the pollen and stigma mature at different times.

There are many dioecious plants in the garden. The winterberry (*Ilex verticillata*) is valued for its brilliant red berries that provide winter interest. Female winterberries will only produce berries if there's a male plant within a 40-foot distance. In most cases you only need one male for several female plants.

Another common plant that is dioecious is marijuana. Since only the female buds are harvested it is important to weed out the males before they can pollinate the female plants.

If you are buying a dioecious seedling you have no idea what the sex will be. To simplify this, nurseries sell cultivars that are either male or female. If you are buying a holly, you can choose between 'Blue Prince' or 'Blue Princess.' I bet you can guess which is the male.

The pawpaw (*Asimina triloba*) is a desirable North American native tree that produces tropical-tasting edible fruit. I grew one of these dioecious plants from seed and now I have to wait to see if it is a female. If it's a male, I will get no fruit. A single female might produce fruit if there is another pawpaw near enough for pollinators to bring pollen. To be on the safe side I now have two more seedlings. Hopefully I end up with at least one of each sex.

Another confounding feature of dioecious plants is how long it takes for the individuals to reach sexual maturity. Male plants grow faster and flower sooner than females.

Some dioecious plants are valued for their shape, foliage or flowers rather than their fruit, so it's not necessary to have both male and female plants. The ginkgo (*Ginkgo biloba*) is planted for its resistance to urban conditions and unique fan-shaped leaves that turn gold in autumn. The fruit stinks—some describe the smell as a mix of rotten milk, dirty socks and vomit. It is a delicacy for some Asian cultures but most people don't want a female ginkgo.

Junipers are another example of dioecious plants that gardeners usually don't need to worry about, since they're valued for their architectural form and evergreen foliage rather than their berries, but some people do harvest the toxic berries.

Jack-in-the-pulpit (*Arisaema triphyllum*) can change its sex. They are usually male at a young age and as they mature and get larger, they become female. Once mature, plants can arbitrarily change sex. After producing fruit one year, females are more likely to be males in the subsequent year.

The jack-in-the-pulpit illustrates the energy cost to a plant to produce ovules and seeds. If the corm is small the plant will probably flower as a male. This is the case for immature plants as well as plants that have had their leaves removed early in the year. Producing pollen takes less energy and allows such plants to grow bigger corms. Once they are large enough, they flower as a female. Deer feed on jacks and in areas where the leaves are normally browsed, you will find few females because they just can't grow big enough.

6
Fruits and Seeds

The end game for flowers is the production of fruit and seeds, both of which are important to gardeners, especially to those producing food. Seeds are also an important source of new plants for gardeners and are really the only way to acquire many of the rarer plants.

Fruits and seeds are a major food source for many animals since they are easier to digest than leaves and stems.

What Is a Fruit?

To a lay person, a fruit is that juicy thing we enjoy eating and includes apples, strawberries and pineapples but in scientific terms a fruit is any plant organ that contains seeds. Beans, nuts and even the dry brown seed heads on your perennials are fruits.

Is a tomato a fruit or vegetable? Most people would call them vegetables, but scientists call them fruits because they contain seeds. Even pine cones and corn on the cob are fruits.

What about maple keys, those winged things that flutter down from maple trees? Those are fruits too and if you break them apart you may find a single seed inside each wing, but they don't always contain seeds.

Only flowering plants produce fruit and it is one of the key reasons they are so dominant and successful.

The Importance of Fruit

One major function of fruit is to protect the seed as it develops and in some cases the fruit protects the seed even after it is fully developed. The fruit tissue also provides nourishment for the developing seed and maintains the ideal humidity. In the early stages of development, fruit is usually hard, green and bitter, and might even contain toxins. In this stage, animals leave the fruit alone.

As seed reaches maturity, the fruit changes. Tomatoes become red, sweet and juicy, advertising to animals that it is ready to be eaten. Many plants rely heavily on the animal's ability to disperse seeds through its digestive system, and as a bonus the seed receives a nice dose of free manure. The digestive acids might even help in softening the seed coat, speeding up the germination process.

Animal-mediated seed dispersal is an excellent strategy for plant reproduction because the new generation of plants can grow farther away from the parent plant, which minimizes competition for resources among the same species. Dispersing the seeds far away also helps plant species colonize new areas, which can be especially useful when the parent plant is harbouring pests or diseases, or the previous area is no longer ideal due to disturbance or environmental change.

Some seeds are too large, hard or bitter to be eaten and are simply left behind by the animal after eating the juicy parts. The walnut is a fairly large green fruit even once it is mature. Squirrels come along and eat the fruit part and leave the seed, or they might just bury the whole fruit. Many of these will be dug up and eaten in winter but some will be lost, allowing them to grow into new trees. I have several large walnut trees which the squirrels enjoy each fall but, unfortunately, I also get a lot of unwanted walnut seedlings.

You may be wondering why some fruit is poisonous since this contradicts the usual mutualistic relationships between fruit-eating animals and plants. There are various reasons for this. The toxin might only be toxic to some animals. The fruit of deadly nightshade is lethal to many mammals but harmless to some birds.

The berries of the high bush cranberry are quite bitter in fall and birds leave them on the tree. Over winter they lose some of the bitterness and birds usually eat them in late winter which may be a better time for seed dispersal. Gardeners take advantage of this since the red berries are quite attractive against white snow.

Spicy peppers contain capsaicin, which is toxic to mammals in high doses and produces a burning sensation when eaten. Birds, on the other hand, have different taste receptors and can't feel the burning sensation from capsaicin. They eat the peppers and fly away, dispersing the seeds over long distances.

Different Types of Fruits

An ovary can contain a single chamber and one ovule, in which case it produces a pitted fruit similar to a peach or olive. It can also contain numerous chambers and ovules which then form a fruit with numerous seeds. Peas and beans have one chamber but many ovules. The apple has several chambers, each containing numerous seeds. Have a closer look next time you eat an apple and you can easily see the chambers.

Raspberries and strawberries are actually clumps of small fruits.

Fruit Development

Fruit development goes through a four-step process: fruit set, cell division, cell expansion and maturation (ripening).

Fruit can be aborted part way through the development process. If energy reserves are not high enough to support all of the fruit, fruit trees will abort some fruit part way through the process. Gardeners also remove some fruit to let the tree focus on producing fewer but larger fruit.

Fruit Set

Fruit set happens when pollination is successful. The plant then has to decide if it will develop the fruit or abort it. Many plants will abort fruit soon after pollination. This includes young plants that don't yet have the energy to develop the fruit properly, recently moved plants and plants that are on the decline. Gardeners should see such fruit loss as a positive thing. The plant is saving its resources for better long-term development.

Fruit set is not fully understood, but does involve a number of hormones and is influenced by the external environment. Temperature plays a big role and both low and high temperatures can result in aborted fruit. Cucumbers can form female flowers and not set fruit. Gardeners may blame this on a lack of pollinators, but it might be due to inappropriate temperatures.

Each plant has an ideal temperature range for fruit set and for many nontropical plants the optimal range is 70°F to 75°F (21°C to 24°C).

Fruit Cell Division

In the cell division phase, cells throughout the fruit start multiplying. This process is highly controlled by the developing seed. There is no point in growing a large fruit if the seeds are not developing correctly, or if there are not enough seeds in the fruit. In tomatoes this phase lasts between seven and ten days.

When cucumber flowers are only partially pollinated so that some ovules are fertilized and others aren't, the fruit may develop curved or have the stem end expand normally while the tip stays small and pointed. The lack of seeds in the tip of the fruit stops proper cell division in that area, hence the constricted size. This is not the only cause of misshapen cucumbers.

The final size of the fruit depends on the number of cells formed during this phase of development.

Fruit Cell Expansion

The previous stage has produced a large number of cells that are not yet fully developed. In phase three, cell expansion, the cells develop further, greatly expanding their size. Both the fruit cells and the cells inside the seeds expand at the same time.

Cell expansion in tomatoes lasts six to seven weeks and some cells will expand in size by as much as 100 times.

Fruit Maturation

At the start of maturation, the fruit has attained its final size and shape. It now undergoes a complex series of changes that includes changes in color, texture, aroma, flavor, sugar content, nutrient content and a drop in toxins. In dry fruits, such as nuts and grains, water is lost and the fruit dries and becomes hard.

Seed undergoes maturation at the same time.

In some fruits this process is managed by ethylene levels produced by the plant. Many people will ripen green tomatoes by putting them into a bag with apples. The ripe apples give off ethylene gas, which speeds up the ripening process in the tomato.

Bananas are picked green and stored in a low-ethylene environment until they reach their destination, which prevents premature ripening. Once at their destination, they are exposed to higher levels of ethylene and they ripen.

When is a tomato fully mature (i.e. ripe)?

We humans and even the birds and animals want our tomatoes to be red and juicy, and we consider this to be the mature stage. From the plant's point of view, the fruit is mature when the seeds are fully developed and able to germinate. That stage in tomatoes is something called the breaker stage. When a tomato fruit has reached the breaker stage, it is still mostly green, but has started getting a pinkish tinge on the blossom end. The seed in such a fruit is fully developed and will germinate.

Gardeners can use the breaker stage to their advantage. If you harvest at this stage, and bring the fruit indoors, it will ripen to a full red color and have as good a flavor as vine-ripened tomatoes. You might not believe that, but a blind taste test will confirm it. The advantage of doing this, is that animals and birds also prefer a red tomato and most will not have taken a bite out of them at the breaker stage.

Seed Development

The development of seed follows fruit development very closely but has only three stages: differentiation, cell expansion and maturation.

Seed Differentiation

Pollination results in the ovules being fertilized with sperm. Each ovule will develop into a single seed.

The single fertilized cell divides, forming more cells and then these cells start to differentiate. This means that they start taking on the characteristics of a complete plant. Some cells start forming a very early form of root cells or stem cells. Specialized cells start forming the cotyledons and the seed coat that will eventually protect the seed after leaving the fruit.

Largely due to cell division, there is a rapid increase in size.

Seed Cell Expansion

The seed continues to grow in size as it forms more cells and existing cells start to expand. A key function at this stage is the accumulation of food reserves. Once a seed is fully developed it not only needs to survive a long time without a food source, but it needs to have enough food stored up to form the initial seedling (roots, stem and leaves).

Seed Maturation

The seed is now fully developed but the cells are too metabolically active and would not survive long without their umbilical connection to the fruit. Cells go into a form of hardening off. They slow down biological activity, while at the same time losing much of their water. The seeds become dry.

The seed coat also gets harder and usually changes color to a dark brown or black.

At the end of this process, many seeds will separate from the fruit. If you go into the garden in fall and shake dry fruits (seed pods), you can hear the loose seeds rattling around inside.

In some cases this phase does not complete properly and the developing seeds start to germinate prematurely in a condition called viviparous germination. This happens in many fruits, but it is reported most often in tomatoes, when someone cuts open a ripe fruit only to find it full of seedlings. These seedlings can be grown into a full plant.

Suckering Tomato Plants

Plants are programmed to produce as many fruits and seeds as possible to sustain their species and tomatoes are no different. If left to their own devices they make large sprawling plants with lots of fruit. This is clearly the best approach for native tomatoes but is it the best way to grow them in the garden?

This leads to an age-old debate. Should you sucker tomatoes?

Suckering is a term used for pruning side branches to keep the plant smaller and more manageable. There are many gardeners on both sides of the fence. Some never sucker and grow their plants in cages to keep them somewhat manageable. Others sucker heavily to a single stem and tie that to stakes, and still others find a middle road, pruning the plant to two or three stems.

Which way is best? That really depends on your goals. Suckering is more work and produces fewer fruits per plant, but it does allow you to plant closer together which results in a higher yield per given space. It also produces an earlier crop which is important in short-season climates. Not suckering produces a higher yield per plant but since the plant needs to support more fruits, they tend to be smaller.

The effects of suckering tomatoes grown three different ways.

	No support, no suckering	Cage, no suckering	Single stem, suckers removed
Maintenance effort	None	Very little	Weekly
Neatness	None	Some	Most
Total yield	High	High	Medium
Number of fruit	High	High	Medium
Fruit size	Medium	Medium	High
Fruit ripening	Later	Later	Early
Disease potential	Higher	Higher	Lower
Slug damage	More likely	More likely	Less likely

Seeds from Non-Flowering Plants

Plants can also produce seeds without flowers and this category includes the cycads and conifers. The seeds they produce have no protective covering and are formed inside cones. This group of plants are called gymnosperms, which means naked seeds.

Maidenhair fern trees (*Ginkgo biloba*) are an ancient conifer that has been around since the time of the dinosaurs. The species is dioecious with male trees producing catkin-like pollen cones and female trees producing pistillate structures containing a single ovule. The seed has a fleshy coating that resembles a small golden plum and is often called a fruit. Seed maturation is usually complete about six to eight weeks after the seeds drop and germination works best after a cold spell.

Soil Seed Bank

Where there is soil, there are seeds. In the 1850s, Charles Darwin discovered that seedlings emerged from soil retrieved from the bottom of a lake. Later, scientists discovered seeds at varying depths within the soil. This natural storage of viable seeds is called the soil seed bank.

Soil seed banks are important because they preserve plant species and genetic diversity the same way human-made seed banks do. In nature, there are various disturbances and sudden environmental shifts that can easily wipe out a generation of plants. If the soil is full of seeds, the ecosystem can regenerate rapidly after a natural disaster or human activity.

One culturally iconic example of the soil seed bank in action is the rapid establishment of the common poppy (*Papaver rhoeas*) after the devastation of World War I. These seeds can stay viable within soil for decades and were some of the first plants to return to the war-torn fields of Europe.

Seeds survive in the soil because they are dormant, an incredible adaptation that ensures seeds won't germinate when conditions are unfavorable. Seedlings have a much higher chance of survival if they germinate when conditions are just right. Seed dormancy also delays germination so that not all the seeds germinate at once. This reduces the chance of a species being wiped out by a severe drought or other catastrophe. Seeds can survive in soil for years and, in rare cases, for thousands of years.

Unfortunately, weeds also produce seeds that go dormant and remain within the soil seed bank. This is often referred to as the weed seed bank. Annual weeds in particular produce large quantities of seed.

Viable seed is lost from the soil seed bank over time. The ones on the surface germinate or are lost to wind, rushing water, predation (eaten by animals or insects), infection by fungi or damage by toxins. All seeds are living organisms with a limited life span and ones lower in the ground slowly die off. Annual and biennial seeds tend to be longer-lived than perennials.

What can you do about the weed seed bank in a garden? Keep soil covered so light does not reach the seeds, thus preventing germination. Over time, the number of seeds is reduced. Stop disturbing the soil. Don't hoe or rototill, which exposes weed seed to light. And lastly, remove all weeds before they can set new seed.

7
The Whole Plant

The previous chapters describe various parts of a plant and now it is time to put them all together and look at the plant as a single organism.

The tissue found in each part of the plant is a bit different, but there are also a lot of similarities. The root, stem and leaves all have an epidermis and this outer layer of cells forms a continuous layer right from the root hairs up to the tip of leaves and flowers. It is like the skin on your body.

All parts of a plant have an interconnected cortex-like center that holds the vascular bundle. The xylem and phloem in the vascular bundle connects every part of the plant together. It is important to think of the plant as one whole system and not a bunch of separate parts.

Life Cycle of Plants

Plants have different life cycles and to simplify things for gardeners they have been divided into different types, including annuals, biennials, perennials, trees and shrubs. This makes it easier for gardeners to understand how a plant grows.

Annuals

Annuals are plants that are genetically programmed to only last one season. Their seed germinates in spring, grows very fast and then flowers. This is a plant on a mission to produce lots of seed, as quickly as possible, so that there will be many future generations.

Annuals tend to make a small root system and don't bother storing any sugars or starches in the roots. Why bother? It won't be around in spring.

The seed produced by these plants tends to be long lived and after a few years the local soil seed bank will be full of them. Many of the common weeds are annuals.

A lot of the annuals grown in temperate regions are not true annuals. In their native habitat they might grow as perennials, but we grow them as annuals because they can't take the cold winters. Cosmos, marigolds and sunflowers are true annuals, but snapdragons, pansies, petunias, begonias and impatiens are really tender perennials.

Gardeners have three options for growing annuals. You can start new seed each year, buy plants each spring or let them self-sow in the garden. I have a couple of different annual poppies that seed themselves around the garden and come up each year. The latter approach is certainly easier, but it provides little control over where they show up, and some annuals can become quite weedy.

Biennials

Biennials are plants that grow in year one and flower in year two. Then they die. Carrots, beets and foxgloves are examples of biennials. Some flowering biennials like foxgloves can return for a third year.

They start out as seeds and usually make a very short plant the first year. Many are simple rosettes of leaves near the ground. They don't waste energy producing taller stems because they don't plan to flower in the first year. The roots are designed to store large amounts of carbohydrates for a quick growth spurt the following spring and they overwinter underground just like a perennial. In the following spring they make a tall stem with a few leaves and then produce their flowers and seeds. They are genetically programmed to die once the seed develops.

Biennials cause gardeners a few problems. The first is that nurseries don't have a biennial section and sell them along with the perennials. Unsuspecting gardeners buy them in full bloom and expect them to come back year after year, but since this is already their second year, they just die out, leaving the customer disappointed and blaming themselves for not growing them properly.

The other problem is in the garden. If you start some from seed you will only have flowers the second year. If you now let them self-seed, there won't be any in bloom for two years. The trick for solving this is to start some in each of the first two years so that seed is produced each year. After a few years you will have flowering plants every year.

Perennials

Perennials are plants that come back year after year. They start life as a seed and during the first year they will usually put all their effort into making a good root system. In some cases you might see flowers the second year, but for many perennials this will only happen in the third year. In some cases like peonies and trilliums, it can take five to seven years before they flower.

Once a perennial flowers it is very likely to flower every year. The clump gets bigger and bigger and produces more and more flowers as it ages. Large mature perennials are a great sight.

There is another class of perennials I like to call short-lived perennials and few gardeners talk about these. Nurseries and books certainly don't advertise them as being short lived. These plants grow for a few years and then die.

There are a couple of reasons for this premature death. The first is genetics. Some perennials are just short lived. North American penstemon are a good example. Even in the wild, many of the species live for two to four years and then die. There is nothing you can do about it—it's genetics. Some penstemon species are good long-lived perennials.

The second reason for short-lived perennials is that they just don't like your growing conditions. It could be the soil, the warm summers, the high humidity or the cold winters. I've tried growing blue poppies several times and have given up. Many of my friends have also tried. They will not grow in our climate. Yet, a few hundred miles away, in a colder climate, they do just fine. I suspect it is our hot, humid summers.

Not all perennials will do well for you, even if the online information says they will. Don't blame yourself for a lack of skill, just accept the fact that some will die.

Grasses, bulbs and woody plants are also perennials, but for some reason the gardening world and nurseries don't consider any of these to be perennials.

Monocarpic Plants

Monocarpic plants can grow for several years until they reach maturity. At that point they flower, set seed and die. This can be quite concerning to gardeners who did not know their plant was monocarpic. Nurseries never tell you these things when you buy them.

Many succulent plants like sempervivums (hens and chicks), echeverias, aeonium, some bamboos and yuccas are monocarpic. Probably the best known is the agave, or century plant. It grows for many years and then one day it produces a monster flower spike and then it dies.

In some cases the plant is completely lost, but in others the mother plant has made some babies to carry on. Hens and chicks do this very well. By the time the mother plant flowers you will have a dozen or more babies.

Here is a trick when buying hens and chicks. Don't buy one large rosette—it might be ready to flower leaving you with nothing. Pick a pot with several smaller plants in it. It is unlikely they will all flower the same year.

Yuccas, like *Yucca filamentosa*, grow a bit differently. They make numerous stems around a central root. When one of the stems flowers, only that stem dies. The others continue to grow and flower in future years.

Why do monocarpic plants die after flowering? The plant's genetics are programmed to transfer all of its food energy from roots and leaves into flowering. There is simply not enough left in the plant to grow again.

You can try to conquer this genetics. In some cases, removing the flower spike as soon as you see it might allow the plant to keep growing. Although this is recommended a lot in social media, in my experience it doesn't work very often.

Some succulents will flower sooner if they are stressed, so growing plants well can delay the inevitable.

Determinate vs. Indeterminate Growth

Do plants keep getting larger? Will a tree keep growing bigger and bigger each year, or does it reach a maximum size?

As discussed above, some plants like annuals and monocarpic plants are genetically programmed to stop growing at some point, and in colder climates perennials stop growing because of environmental conditions. Many other plants keep getting larger year after year.

A determinate plant is "determined to stay small." That is not an official definition but it helps me to remember the meaning of the term; just keep in mind that "small" is relative. Determinate plants grow to a limited size and then stop getting bigger.

Indeterminate plants keep growing larger their whole life. Their genetics does not determine their ultimate size.

Most deciduous trees are determinate. A redbud grows to about 25 feet in height and then stops getting taller. Sugar maples grow to about 70 feet. When you purchase such a tree, the plant tag will give you the maximum height.

Evergreens on the other hand tend to keep growing bigger their entire life, although the rate of growth does slow down as they get older. Some plant tags will try to give you the maximum height but that is not very meaningful since they are indeterminate. Better quality nurseries will give you the ten-year height and say something like the height is 20 feet (ten years). This is the height of the plant after growing ten years. You can take the given height and divide by ten to get the annual growth rate which allows you to buy evergreens for the space you have. My slowest growing evergreen is a spruce that grows ¼" per year.

The terms determinate and indeterminate are also used to describe some vegetable crops such as tomatoes. A determinate tomato cultivar will grow to a specified height, produce fruit and stop growing. Indeterminate tomatoes are vines that just keep getting taller and produce fruit until winter kills them.

Other crops that are available as both determinate and indeterminate cultivars include beans, peas, and cucumbers. Determinate plants, also called bush types, are a better choice for growing in containers, since they stay smaller. They also tend to produce fruit earlier in the season.

Indeterminate plants tend to have longer days-to-maturity but they produce food over a longer period of time. They do need more space.

Gardeners talk about determinate and indeterminate potatoes, but they don't exist. Potato plants are determinate and it is better to use the terms early-season, mid-season and late-season to describe how and when they produce tubers.

Plant Dormancy

Some plants go dormant for a while but there is always some metabolic activity, so they are not truly dormant. When gardeners talk about dormant plants they are usually referring to the visible top growth. So a plant is considered dormant if there is no new green growth. Tulips are considered dormant all summer and fall even though they are making roots in late summer and fall. Shrubs are dormant in winter even though the roots might be actively growing.

What causes this dormancy?

Cool-growing lawn grasses grow best in cool temperatures. As things get hot in midsummer they go dormant, only to begin growing again in fall when it cools down. This is why you should fertilize mostly in spring and a bit in fall.

Water also plays a role in dormancy with cool-growing grasses. If you keep watering the lawn all summer long it tends to stay green. It still grows less in summer heat, but water can overcome the warm dormancy period to some extent.

Ornamental grasses come in cool growers and warm growers. The blue festuca is a cool grower and starts to grow early in the year. Miscanthus are warm growers and just sit there in spring as brown stubble until things get warm enough. It is important to take this into account when designing beds. Miscanthus looks better near the back of the border so you don't notice its ugly spring phase.

Plants native to temperate regions tend to go dormant based on night length and temperatures. Desert and tropical plants are controlled more by water cycles.

Water is critical for plants and a drought condition will force plants to go dormant. When things get too dry, root hairs are lost and not replaced. This dramatically reduces the water reaching leaves which start shutting down. If the drought continues, leaves will also be abandoned.

Many deciduous plants go dormant to spend the winter without leaves. It is generally thought that this process starts in fall, but it really starts in mid-summer. It is a complex process that takes many weeks and the increasing night length signals plants to start preparing for winter.

Movement of Water

Plants are 95% water and water impacts every aspect of a plant's life. Water is absorbed through the roots where it enters the xylem. From there it flows up through the stem and along branches until it reaches fruits, flowers or leaves. The majority of the water ends up in the leaves.

This movement of water brings nutrients from the roots to all parts of the plant. Water fills cells which become turgid and provide support for the plant. As water evaporates out of the leaves it cools the plant. Water really is critical to plant growth.

As leaves lose water to transpiration, the amount of water in the leaves drops which causes a suction force on the xylem. The suction force pulls more water up from the roots.

You may be wondering how it's possible for water in a plant to move upwards against gravity. It is largely due to the structure of water molecules. A water molecule consists of an oxygen end which is negatively charged and two hydrogen ends which are positively charged. This causes the water molecule to act like a small magnet. Think of a string of magnets, one end attached to the other. If you pick up one magnet, you pick them all up. Water behaves the same way.

Each water molecule sticks to the end of other water molecule above it, forming a long chain. You can see the effect of this easily by looking at a drop of water. It does not flatten out, but forms a round ball. This is due to the fact that each water molecule sticks to others and a sphere is the most efficient shape to hold a lot of molecules.

Water Moving Between Cells

The previous section describes how water moves up a plant but it does not describe how it moves through a plant once it leaves the xylem or even how it gets out of the xylem.

There is a second force at play called osmosis. Water naturally moves from an area of high concentration to an area of low concentration. The xylem is full of water, a high concentration, so it naturally flows out of the xylem to neighboring cells that have less water, a low concentration. As cells next to the xylem get full of water, it starts to flow to more distant cells that have even less water. Water flows between cells in the direction of lower concentration. In this way water gets to all parts of the plant.

Guttation

When stomata are closed or in conditions of high humidity it can be difficult for a plant to get rid of excess water. In such cases roots can absorb too much water and the excess still flows up the plant and into cells. When this happens, the excess has to go somewhere so it leaks out of small channels in the epidermis, forming water droplets on leaves. You might notice this on lawn grass where water droplets form on leaf margins in the morning after conditions of low evaporation.

Movement of Nutrients

Nutrients are normally absorbed by roots, and travel through the xylem along with water to other parts of the plant. They can also be absorbed through the leaves and stems, either naturally, or by foliar sprays (discussed below).

The flow of nutrients is not just random and to some extent plants can control where nutrients end up. They tend to supply more nutrients to younger leaves than older ones because that is where they are needed most. This explains why a soil nutrient deficiency shows up in older leaves first—they are cut off if there is a shortage.

Nutrients that are absorbed by roots move up the plant using the xylem. Once at their destination they might remain there or the plant moves them to other locations where they are needed, and this happens using the phloem. But there is a catch here. Not all nutrients can enter the phloem. Nutrients are divided into two groups: mobile nutrients and immobile nutrients.

Small molecules or ones with a smaller positive charge move into the phloem easily and are able to be transported to other parts of the plant. These include nutrients like ammonium, potassium, phosphate and magnesium, as well as urea (not a plant nutrient).

Larger molecules and those with a stronger positive charge do not move freely throughout the plant and can't enter the phloem. These "immobile nutrients" include calcium, iron, manganese, zinc, boron, sulfur and copper. These nutrients remain in their current location and can't move to other parts of the plant.

This movement of nutrients is important to plants. In the fall, plants move mobile nutrients from leaves into woody stems and roots resulting in yellow leaves.

You might have noticed that older leaves on growing perennials turn yellow as they age. The plant is aborting these old leaves which have done their job and are no longer needed. Before dropping them, the plant moves nutrients out of the leaves. A plant that is deficient in nitrogen will also move nitrogen out of lower leaves and move it to new leaves. The yellow lower leaves may be telling the gardener that there is a nitrogen deficiency.

> ### Plant Myth: Leaves Can Be Used to ID Nutrient Deficiencies
>
> The internet is full of memes that show you how to identify soil deficiencies simply by looking at leaves. None of these work because plant biology is far too complicated.
>
> Here is a common example: chlorosis indicates iron deficiency. A plant is reported to have chlorosis, a yellowing of the leaf, and the diagnosis is iron deficiency.
>
> Firstly, iron does not cause chlorosis of the leaf. It actually causes interveinal chlorosis which is the yellowing of the spaces between the veins of the leaf.
>
> The second mistake is assuming that there is only one cause for this symptom. Numerous conditions can cause interveinal chlorosis, including:
>
> - manganese deficiency
> - a high soil pH
> - zinc deficiency
> - herbicide damage
> - wet soil conditions
> - compacted soil
> - trunk-girdling roots
> - plant competition
> - high organic content in soil
> - high salts
> - iron deficiency
> - high levels of phosphorus, copper, zinc or manganese
>
> The bottom line is that interveinal chlorosis tells you there is a problem but it does not indicate the reason for the problem.

Here are some general rules to help you figure out nutrient deficiencies:
- Mobile nutrients move to areas of active growth; immobile nutrients do not redistribute within a plant.
- Mobile nutrients move in all directions: up, down and sideways; immobile nutrients stay in one place.
- Deficiency symptoms of mobile nutrients appear first in older leaves, and if not corrected they show themselves in new growth as well. A deficiency of immobile nutrients will show up in new growth first.

Foliar Feeding

Foliar feeding is a method of applying nutrients directly to leaves instead of the soil. This seems to make a lot of sense—get the nutrients where they are needed using a more direct approach. This method of feeding plants is becoming more popular, mostly due to strong advertising of less than truthful benefits.

How do the nutrients get into the leaf? Many incorrectly believe that the sprayed nutrients enter the stomata. Research has shown that this is not what happens because stomatal guard cells are coated in waxy material which keeps water out.

Most of the absorbed nutrients from foliar sprays enter the plant through transcuticular pores. These are extremely small openings in the cuticle and epidermis through which small molecules, like nutrients, can enter the leaf. For this to work the molecules have to be small, so iron ions do get in, but chelated iron does not.

It is true that adding nutrients to soil encounters significant losses due to leaching, chemical reactions and use by microbes. However, the leaf can only absorb small amounts of nutrients. Soybeans remove 3.5 pounds of nitrogen, 0.8 pounds of phosphorus, and 1.4 pounds of potassium per bushel of seed produced. Foliage fertilization could never provide these levels of nutrients. Roots are much better at absorbing large quantities of nutrients.

Only about 15 to 20 percent of the nutrients applied to leaves are absorbed. Foliar feeding is best used as a fast short-term fix for a nutrient deficiency and not as a long-term feeding strategy. In my opinion, gardeners should stay away from such products.

I routinely see suggestions for spraying tomato plants with a calcium foliar spray to prevent blossom end rot (BER). There are two problems with this suggestion. First, BER is not due to a calcium deficiency—it is a watering issue. And secondly, the calcium absorbed by leaves will stay in the leaf. It won't move to the fruit because it's immobile. Calcium in the fruit must come from the roots via the xylem. Partially developed tomato fruit does not absorb calcium through the skin.

Movement of Sugars

Sugars are produced in leaves and stems and are needed in all parts of the plant as an energy source. They are also stored in underground root structures during dormant periods. It is important for the plant to be able to move these to areas that need them.

Different areas of a plant are either a source of sugars or they need sugars, in which case they are called a sink. Mature leaves and stems are sources. Roots are usually a sink because they can't produce their own sugars. Newly forming leaves in spring are sinks for sugar until they get large enough to produce their own. This also applies to new growing shoots. They start out being a sink and once they develop enough leaf area they become a source.

This sink-source relationship can change with the seasons. In fall, roots are a sink and sugar is sent to them for storage, but in spring roots are a source providing sugars to terminal buds. The root can also be a source in the middle of summer if top growth is removed. This explains why the taproot of a dandelion starts growing after removing all its leaves.

Sugars move from one part of the plant to another via the phloem. Consider a leaf in the middle of a plant. It is a source and makes sugars. Now these sugars need to move down to the roots to support new root growth, while at the same time moving up to the growing terminal bud. All of this happens with the phloem.

The secret lies in the design of the phloem. It is made up of a number of different tubes, like a handful of drinking straws. At any time, some are moving sugars up and others are moving sugars down the plant. The number of up and down tubes is not static. In spring, most phloem tubes are moving sugars up the plant to support all that new growth. In fall most tubes are moving sugars down, for winter storage.

What about that leaf in the middle of the plant? Its sugars will go in whatever direction the plant needs and they might even travel in both directions at the same time. The sugars from leaves near the growing tip tend to move up and lower leaves send sugars down. Flowers and fruits get most of their sugar from the closest leaves.

I have described how sugar moves in the phloem, but this description also applies to many other molecules including mobile nutrients, proteins and root exudates. They are able to move around the plant to where they are needed most.

Seasonal Sharing of Resources

Different parts of a plant are good at sharing resources. Roots grow when it is cooler—fall and late winter. Shoots and leaves grow in spring and summer. In this way the plant spreads out its use of resources over the whole year.

Gardeners want to plant in spring and right after planting the priority for plants is to grow roots, which now has to happen at the same time as growing shoots and leaves, putting a huge strain on plant resources. That's why the earlier in spring you plant, the better it is for the plant.

This also explains why fall planting works well. The top growth no longer needs resources and may even be sending them down to the roots. The season is getting cooler, with more moisture in soil making this a great time for planting and root growth. For this to work well it is important to keep watering right up to the point where the ground is frozen.

Overcoming Physical Damage

When an animal gets hurt, the injury heals over time as the damaged tissues are replaced by fresh new cells. Plants can't heal injured tissues in the same way but they are able to correct many forms of damage and in some ways their ability to recover from damage is superior to animals'.

If an insect takes a bite out of a leaf, the unharmed cells around the damage form a kind of wall to separate good cells from damaged cells. Remember that all of the cells in a leaf are connected. Water, nutrients and sugars easily flow from one cell to another. Right after the damage these important compounds start leaking out of the leaf, so it responds by walling off the damage to stop the leakage and prevent pathogens from entering the leaf.

You can see this wall in some types of damage as brown tissue at the edge of good cells.

Once the wall is in place, the leaf goes on to function normally. If you go out in the fall and look at leaves that have holes in them you'll see that the remaining parts of the leaf are fine. Metabolically, it is business as usual.

The same mechanism is used if the break occurs on roots, the stem, or on a tree branch. A wall is formed between good tissue and damaged tissue. Eventually the damaged tissue falls off.

Once the damage is no longer a threat to the whole plant, the plant can begin new growth, which happens in most parts of the plant except in leaves. A plant may make more leaves but it won't repair the damaged ones.

The plant makes use of existing meristems which start rapid cell division, followed by differentiation. If a stem or branch breaks the new growth usually starts from a dormant bud below the break. If a root is damaged the plant may just wall it off and use existing root tips to grow more roots. Or it might start forming new meristematic cells at the point of injury and form a whole new root.

Many plants have hidden dormant buds that you can't see. If you cut a sugar maple tree to the ground, new dormant buds below the ground start to grow to form a new tree. Do the same with a spruce and the tree dies because it does not have dormant buds near the ground.

You have probably tried to dig out a dandelion. If you cut off all the leaves along with about one inch of root, it will regrow by forming a new shoot. If you cut the leaves and three inches of root, it usually dies.

This ability of plants to endure significant damage and keep on growing is one reason that they are so successful, but it should also tell the gardener something. Most damage to a plant won't kill it. Learn to live with a few rabbit bites and holes in the leaves. A tent caterpillar nest on a healthy shrub will not do much harm.

How Do Plants Get Taller?

Most plant growth happens in shoot tips and root tips. They both contain meristematic cells that are rapidly multiplying. As they differentiate and become mature, their size becomes fixed—they can't grow anymore. So how does a plant get taller?

Since the cells in a mature stem are a fixed size and they don't multiply, the plant can only grow taller at the shoot. It grows by continually pushing the tip of the shoot higher and higher. The same thing happens in the roots. Only the root tips grow in length. Older root tissue can grow in width, but not in length.

This also applies to trees. They only get taller at the shoot tip in the terminal bud. If you place a nail in a young tree at six feet above the ground, it will still be at six feet when the tree is old. The trunk expands sideways and gets thicker, but the trunk does not grow in length.

Following the Sun

At daybreak the sunflower faces east. As the sun rises and moves across the sky the sunflower follows the sun. By the end of the day it is facing west and by next morning it faces east again, ready for another day.

Contrary to popular belief it is only the immature flower that does this. Once the flower has set seed it stops following the sun and remains facing east.

Why does the sunflower do this? It turns out that facing the sun first thing in the morning helps warm up the flower and a warm flower gets more pollinators visiting than a cool flower. More pollinators translates into higher seed production.

The flower starts following the sun when it is still an immature flower showing only green sepals. All this green tissue is photosynthetic and helps produce food. Facing the sun makes it more productive.

An even more interesting question is, how does it follow the sun? Some recent research has shown that the immature flower stem elongates faster on the sunny side, so one side gets longer than the other, which twists the flower head as the sun moves east to west. Once the stem is fully developed, it can no longer twist.

The sunflower is not the only plant that does this. Many alpines also turn their flowers as the sun migrates across the sky. At high elevations warmth can be critical for both pollinators and for maturing pollen.

Many plants also turn their leaves throughout the day to maximize the amount of sun they capture.

How Light Affects Plant Growth

Plants that receive plenty of light can carry out all of the processes discussed in this book, which results in healthy, flowering plants. As the light level drops, plants are unable to do it all and they use their food in a hierarchy. The top priorities are core functions such as photosynthesis, respiration, movement of water and nutrients, and pest control. Growth is the next level of priority and the lowest priority is reproduction.

As light levels drop below optimum the first thing that is stopped is reproduction. That is why many sun-loving plants grow well in shade but reduce or even stop flowering. Even shade-loving plants need a certain level of light, or they stop flowering. Low light may explain why a purchased flowering plant never flowers again. It was probably grown in a lot of light to get it ready for sale and now it is sitting in a shady corner.

In low light plants bend and grow towards the light which usually results in crooked plants. They will forgo growth in width and grow tall instead, producing weak plants. Spindly seedlings are a good example of this.

As light levels drop further, the plant stops growing. Plants that are put in too much shade continue to live but they never seem to get bigger.

Drop the light even more and plants start becoming unhealthy. They can no longer maintain basic metabolism and are much more likely to succumb to pests and diseases.

Moving weedy sun-loving plants into more shade slows down their spread. Many houseplants growing in the low light levels of our home don't flower well but a summer outdoors can dramatically increase their flowering.

Nothing, including fertilizer and water, can compensate for a lack of light.

Gravity

Plants know up from down. As a seed germinates, the roots grow down and the shoot grows up, even if the seed is upside down. The same goes for bulbs—they don't need to be planted the right way up to grow properly.

What is more interesting is that they also know when branches are perpendicular to the ground. We know where our body parts are in relation to each other by something called proprioception. It is our sixth sense (not ESP as many believe) and it allows us to move our legs in a coordinated fashion to walk or swing an arm to catch a baseball. Plants have a similar mechanism that allows them to know where their parts are and their relative position to one another. This explains why a plant always has the same shape. Some are round. Some are more vertical. This is very evident in tree shapes.

A few years ago upside-down tomato plants became popular. Tomatoes were planted so they grew out of the bottom holes of pots and people claimed it made for easier picking. Anyone who understood that plants sense gravity knew right away that this would never work. Roots grow down and stems grow up and the poor plant tries to right itself. The whole thing was a silly idea.

8

Woody Plants

Woody plants include most trees and shrubs as well as some so-called perennials that are really small shrubs, like lavender. They are commonly called "woodies."

The focus so far in this book has been on herbaceous plants and most of that information also applies to woody plants but this group of plants also has some special characteristics that are discussed in this chapter.

What Are Woody Plants?

Trees and shrubs are easily recognizable woody plants, but what makes them woody? What is the difference between a perennial and a woody plant?

Woodies are plants that produce wood as their structural tissue and thus have a hard stem.

Herbaceous plants retreat underground for the winter or dry dormant season while woody plants maintain at least some of the aboveground growth throughout the dormant season. New growth starts from existing woody stems in spring allowing them to get bigger each year.

Trees and shrubs are well known but there are other types of woody plants. Many vines are woody. The grape vine, for example, produces a woody stem and new growth starts from there. Clematis also forms woody stems that survive the winter, but gardeners might cut these back to the ground depending on the flower type.

A group of woody plants that are not well recognized are the subshrubs. A subshrub is really just a small shrub but their stems are so short that the gardening industry calls them perennials. These include things like lavender, Russian sage, culinary sage, thyme and heather.

The trick to growing subshrubs is to treat them like larger shrubs instead of perennials. Prune like a shrub and don't cut them to the ground like a perennial.

Structure of Woody Stems

An older woody stem has a central heartwood, sapwood, a cambium layer, inner bark and the outer bark.

On thick wood, most of the internal space is made up of heartwood. This is old xylem that is now clogged with tannins and resin and is no longer able to move fluids. It does add important structure to the tree.

The sapwood is also made up of xylem tubes but it is still functional. The innermost layers are the oldest and will slowly be converted to heartwood once they are no longer needed. The outermost layer of sapwood is young xylem that remains functional for several years.

The annual xylem growth is what forms the rings in wood. Each layer is one year's growth.

Next is the cambium layer which is the growing part of the branch. It contains meristematic cells that grow into xylem tubes on its inner surface and phloem tubes on its outer surface. By growing new xylem and phloem cells it not only increases the number of these cells, but it also increases the thickness of the branch.

The inner bark is the active phloem. The cambium layer keeps making more phloem cells and the older ones die off and become part of the outer bark.

The outer bark is made up of dead cork cells designed to protect the inner bark. Mature cork cells are dead and primarily made up of a waxy substance called suberin. Suberin is impermeable to gases

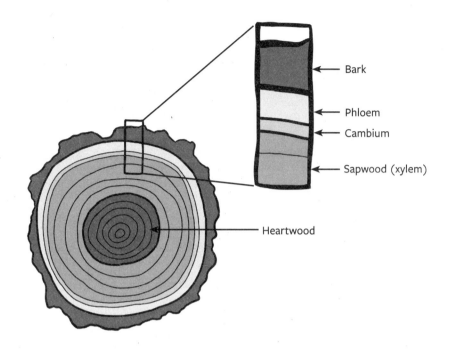

Cross section of a branch.

and liquids and helps the cork cells insulate and protect the woody stem. Cork cells are usually hollow when they are mature, and they can contain substances like tannins that further protect the bark from predation.

Lenticels

Lenticels are special openings in the bark that allow gas exchange between the inner parts of the stem and the air. They serve the same purpose as stomata on leaves. Oxygen is taken in for metabolism and excess water and CO_2 are given off.

These can be quite small and invisible to the naked eye or very distinct. The dark horizontal marks on birch and cherry trees are lenticels. They also show as tiny white or dark marks on fruit such as apples and pears.

Cherry tree showing horizontal lenticels.

Softwood vs. Hardwood

Softwood and hardwood have a couple of different meanings when we are talking about trees. One use is to describe the quality of the wood. Hardwood is a better quality of wood for many applications like furniture. The wood has more cells per inch and less air than softwood, making it harder. Ash and beech are both hardwoods.

These terms are also used to describe the hardening off process that takes place on first-year woody growth. New growth is soft and green and is called softwood. In many cases it is so soft you can take the end of a new stem and twist it around to form a knot.

As it matures it gets harder and darker. It still bends but not nearly as much and is now called semi-hardwood.

Maturation continues as internal wood and external bark is formed. This is now hardwood.

We think of softwood as existing in early spring, semi-hardwood in mid-summer and hardwood in fall, and to some extent this is true, but it is more complicated than that. The maturation process is continuing all the time. If you look at a new stem in mid-summer you will find that the young tip is softwood, but the lower section, which is now several weeks old, is already semi-hardwood. This hardening off process is gradual and happens all along a stem at the same time.

Where Does Wood Come From?

The above description of woody plants is superficial and does not explain why woodies have wood. Why does wood form on trees and not perennials? What is the process for forming this wood? To answer these questions it is important to understand secondary growth.

Primary and Secondary Growth

All of the growth discussed so far in this book is primary growth. Buds expand and develop into stems, leaves and flowers. Stems and roots elongate. It is called primary growth because it produces the main organs.

In herbaceous plants that is the end of growth. Once a stem reaches its full thickness, growth stops. The same is true for leaves, flowers and fruits.

Woody plants don't stop there. They have the ability to carry out secondary growth on their stems and roots, but not on their leaves, flowers or fruits. This secondary growth allows the plant to continue to expand in width over many years. It also allows the added growth of a bark layer which is more prominent on stems than roots. Tree trunks are just old stems.

The secret to this lateral growth is the fact that woody plants have meristematic tissue, called lateral meristems, right in their stems and roots. As these organs grow they activate this meristematic tissue, which starts to multiply forming many new cells. This is followed by differentiation producing more epidermis, cortex, xylem and phloem cells. As this proceeds, stems and roots get thicker each year.

Woody plants also have specialized secondary xylem and phloem cells. Secondary xylem cells differentiate into wood that eventually forms most of the inner part of a branch. Secondary phloem cells differentiate into inner bark. The outer layer of this becomes the visible bark on the outside of trees. It becomes thicker and thicker over time.

You can easily see the changes in bark by comparing one-year-old stems with the trunk of the same tree. New stems have no bark

Plant Myth: Newly Planted Trees Need to Be Staked

Most trees do not need to be staked after planting. The exceptions are very tall plants with very small root systems, or locations that are very windy. I have planted a lot of trees and staked hardly any. Even when staked, it should be loose so the tree can wiggle, which encourages proper root growth. All stakes should be removed after a year.

Here are some other myths about planting trees:
- Dig a deep hole. Wrong, the hole should be no deeper than needed so that you end up with the root flare at ground level.
- Amend the soil. You should not amend the soil. The tree will be making very wide roots and it needs to be able to grow in your soil. An amended hole can cause watering issues—don't do it.
- Prune the top back to match the loss of roots. This sounds like it makes sense, but it is not a good idea. Pruning encourages more top growth and right now you want lots of root growth instead. If the tree can't support all of the top growth it will decide to abandon leaves and drop them.

and the outer layer consists of mostly a thin epidermis. As that stem ages the epidermis is replaced by a harder, darker skin that is the start of a bark layer. In old age the bark can be more than an inch thick.

The type of bark and thickness depends very much on the species. Oaks have thick dark bark, while beech trees have a thin gray bark. Birch trees are known for their white bark.

Storage of Sugars

The classical explanation of sugar storage in trees goes something like this: In fall all of the extra sugar in the leaves is sent down to the roots for winter storage. In spring, it is again pumped up the tree to where the new leaves are forming. In sugar maples you can see this as maple syrup rising up the tree.

This explanation is not completely correct. If it were, you would expect buds near the bottom of the tree to grow first, while the upper ones wait for the sugars to get there.

It is true that some sugars are sent to the roots for winter storage, and some does move back up the tree in spring, but a lot of the sugars stay in the upper part of the tree all winter. Some stays right in the dormant buds where it acts like an antifreeze, protecting the buds in winter.

A lot of the sugars are converted to starch and stored in sapwood throughout the tree. It is now close to dormant buds, allowing all buds on a large tree to leaf out at the same time.

Taproots vs. Fibrous Roots

Roots of dicots, including most shrubs and trees, make an initial taproot. You can easily see this when a seed germinates. In most perennials the plant quickly transforms this into lateral roots forming a fibrous root system.

A common belief is that some trees continue growing long taproots and others form fibrous root systems. You can easily find lists of tap-rooted trees on the internet which include things like oaks,

Taproot on an ash tree seedling.

pines and walnuts. It is true that these trees form a large taproot in the early years of growth but they soon start growing a fibrous root system and it is these roots that are responsible for most of the trees' growth.

The reason for making a taproot initially is for support. It is a good way to provide quick support for a tall tree. The above picture shows an ash tree seedling with a thick taproot and smaller, more recently developed, side roots.

The problem with taproots is that they are not very good at providing air, water and nutrients. Most of the air in soil is found in the top few inches of soil. This is also where you find most of the organic matter and most of the microbe activity. Both of these provide much of the nutrients a plant needs.

Over time the initial taproot is either absorbed or, if it remains, it loses most of its functionality except for support.

The depth of the fibrous root does vary depending on species and soil type, but most feeder roots are in the top foot of soil.

Composition of Wood

Ever wonder what a tree is made of? Gardeners talk a lot about nutrients so one might expect that it contains a lot of nitrogen, phosphorus and potassium.

A very elegant experiment was carried out in the 17th century by Jan van Helmont. He placed a five-pound (2.2 kilogram) tree in a pot containing 200 pounds (90 kilograms) of dry soil. For five years he only added rainwater. At the end of the experiment, the tree weighed 169 pounds (77 kilograms) while the soil lost two ounces (57 grams). Where did all the weight come from?

The weight gain is due to CO_2 and H_2O. Nutrients add almost nothing to the weight of a tree. Dry wood is composed of 50% carbon, 6% hydrogen and 44% oxygen. Everything else adds up to less than 1%.

This clearly illustrates the value of trees for removing CO_2 from the air in an effort to slow down global warming.

Apical Dominance

Apical dominance is present in all plants, but it is much stronger in woody plants and it is critical to understand it when pruning woodies.

If you look at a dormant tree branch you will clearly see a bud at the tip of the branch. This is the apical, or terminal, bud. Below that you will see more buds: the lateral or axillary buds. In most cases these will be smaller than the apical bud.

I like to describe apical dominance this way: The apical bud is the king bud. The king wants to rule all of the other buds and limit their growth. So the king produces a hormone called auxin and sends it down the phloem to all the other buds. Auxin has a special power—it prevents the other buds from growing.

In spring, the king bud starts to grow and keeps sending auxin down the stem preventing other buds from growing. The plant sends all of its sugar resources to the king and it grows quickly. This is apical dominance. The apical bud dominates the other buds and keeps them from growing.

Upward growth is critical for most plants to compete for sunlight. Apical dominance allows the top of the plant to grow higher as quickly as possible without wasting resources on side branches. It also results in the formation of a single leader which, in old age, produces a single strong trunk.

The effect of auxin is strongest right below the apical bud and gets weaker as you move away from it. This allows lower buds to also grow and form side branches.

The effect of auxin is positional: it tends to flow down. If you take the leader and artificially bend it down so that the apical bud is below other buds on the stem, the ones that are physically higher up will be released from apical control and start to grow. This is commonly used in espalier training. It also explains part of the growth seen in weeping trees where a stem starts growing up and then curves over to grow down. This releases the uppermost buds from control and they grow to form new branches.

Damage to the Apical Bud

What happens if the apical bud is damaged? If the central leader of a tree is broken off, say in a windstorm, the apical bud is also lost. Without it, no auxin is produced and the buds immediately below the break start growing.

They all want to become king buds and try to hog the plant's resources. They also produce auxin, which controls lower buds but not the growing ones. This results in several new leaders.

In nature one of these might become dominant over the others and the tree regains a single leader. What is more common is that more than one keeps growing, resulting in a tree with several trunks. Such trunks grow fine for a while, but as they get thicker they form a very narrow crotch between them, which is very weak. At some point, it is very likely that the two trunks will split, causing major damage to the tree.

In a garden, this type of growth should be controlled as soon as it happens. Remove all but one leader, and restore a single king bud.

Espalier trees are created by controlling apical bud dominance.

Pruning the Apical Bud

Gardeners prune out the apical bud to form all kinds of interesting trees.

Coppicing is a technique whereby the apical bud, or even the whole main trunk, is removed to encourage many new growths. The resulting tree looks more like a shrub with many branches. This is a common way of producing willow stems for weaving projects and wattle fences. In the past it was also used to produce firewood.

Pollarding is a similar practice of removing upper stems to increase the density of the branches. It is also a way to keep trees smaller.

Espalier fruit trees are also produced by exploiting apical dominance. These trees are initially trained to grow vertically, then trained to grow horizontally so that the tree can produce relatively high quantities of fruit without taking up a lot of space. The tree can also be manipulated to absorb as much heat and sunlight as possible.

Pruning hedges and topiary also removes apical buds resulting in more side branch growth and fuller trees or shrubs.

Epicormic Buds

Epicormic buds are buds found along a woody stem that lie dormant beneath the bark. Hormones from active shoots completely suppress their growth and they can remain undeveloped for a tree's entire life. They are common in deciduous plants and a few evergreen species.

Stress stimulates the growth of epicormic buds, including sudden environmental changes, thinning, crown dieback, heavy pruning, root death, cold or change in the water table. Producing these buds is an excellent survival strategy for trees that live in ecosystems with frequent disturbances. It's like taking out life insurance.

Many trees can be cut down to a stump and still grow new branches since the epicormic buds beneath the remaining bark start developing. In our area the ash trees seemed to have been killed by the ash borer, but these trees sprouted at ground level. These buds are also sometimes activated after conditions become more favorable or less competitive, such as when another tree dies, opening up the canopy to more light.

Some trees produce water sprouts, which are characterized as fast-growing vertical branches, usually growing out of trunks or large branches. These develop from apical buds. Fruit trees tend to grow water sprouts because they are heavily pruned. Water sprouts should normally be pruned out.

Gardeners make use of these kinds of buds when using pruning techniques like coppicing and pollarding.

Epicormic buds are present in many woody plant species, though they are absent in most coniferous trees and shrubs. Therefore pruning needle-leaved species can be trickier. If you make a mistake pruning a coniferous tree, such as pruning a branch back too far or pruning too many branches off, new branches won't grow from the stem to hide the mistake.

Dormant Buds

This is more of a gardening than a botanical term. Gardeners use the term "dormant bud" to refer to any bud that has completed early development and then stopped growing. These buds are not dead even though they might look dead.

Such dormancy can occur when temperatures are too low or if water is scarce. It can also be a normal part of the plant's cycle. Many woody plants produce leaf and flower buds in late summer and fall, which then go dormant until spring.

The term is also used for perennials. For example, peonies form dormant buds (or eyes) underground which start to grow in spring.

Gardeners talk about dormant buds on roots forming suckers. In most cases these are epicormic buds. I have used the term dormant buds in this book in a more general way, referring to both regular buds as well as epicormic buds.

Healing Damage

The discussion in the section called "Overcoming Physical Damage" also applies to woody plants but there are some additional topics worth discussing.

As branches grow they develop special tissue at the point where the branch connects to the trunk or a larger branch. It appears as a swelling of the trunk around the branch and is called the branch collar.

When the branch is removed either by pruning or natural dieback, the tree will seal off the damage and then grow the branch collar to cover the cut. In time, this growth will completely seal off the area and start growing true bark tissue. It can take years for a large cut to seal off completely, but at some point you will hardly see any evidence of the previous branch.

Branch collar develops as a bulge around the branch, the white line shows the right place to cut off the branch

New wood starting to grow over the cut branch

Years later and the cut is almost fully covered

Branch collar grows over a removed branch if the branch is removed correctly.

Plant Myth: Damage on Trees Should Be Painted

Many gardeners paint cuts and damage on trees with latex paint or apply various commercial tar products that are promoted for this purpose, but this does more harm than good. A covering seals in moisture allowing microbes to grow and further damage good wood.

The best thing to do for newly pruned wood, or damaged trees and shrubs, is to leave them alone. They will take care of themselves.

When you are removing larger branches, it is important that you do not damage the branch collar. If it's damaged too much, it won't grow.

Girdled Trunks

The trunks of young trees have very thin bark to protect them. The phloem and cambium layer are right under the surface and the phloem usually contains sugars. In winter it is quite common to have small animals like rabbits chew off this outer layer of bark, resulting in a girdled trunk.

Without a cambium layer, the tree can't make new phloem cells and it dies. Trees that have been girdled all the way around will lose their upper growth but may resprout from below the damage. A tree with at least half of the bark intact will likely survive but will be weak for a number of years.

Conifers

Coniferous and deciduous trees are fairly similar, even though they look very different. Conifers have needles instead of leaves and most of them do not drop their needles on an annual basis. The structure of the xylem is slightly different, but it still transports water and nutrients up the tree. Most of the topics in this book also apply to conifers, but there are some differences worth discussing.

Dormant Buds on Conifers

Conifers form dormant buds just like deciduous trees, but there are fewer of them and they don't live very long.

Dormant buds are formed at the tip of branches in one year and some of these will grow the following year. As long as there are needles on a branch, the buds on it tend to be alive. By the time needles are lost on a section of branch, the dormant buds are usually dead.

What this means is that if you prune a branch back to a point where there are no needles, it won't grow because all of the buds are dead.

Most conifers also have very few if any epicormic buds. If a pine tree is cut down it won't produce suckers or even a new shoot from the remaining trunk because there are no epicormic buds to grow.

There are exceptions, yews being one of them. They have epicormic buds throughout their woody branches and trunks. Cut them to the ground and they will make all kinds of new shoots.

Needle Drop

It is a common belief that conifers keep their needles forever, but that is not true. Most trees, like pines and spruce, will keep them for a few years and then start dropping them. If you look at the inside of these trees you will see branches that are completely bare.

Older needles are dropped for a couple of reasons. They stop being as functional as new needles but the real reason is that new growth at the tip of branches causes too much shade on old needles, so there is no point in keeping them.

Metabolism in Winter

Confers need to survive winters and unlike deciduous plants, their needles remain green and alive all winter long. They may look dormant but they are metabolically active all through winter. They photosynthesize, absorb CO_2 and release water through their stomata. This process slows down as the temperature drops and light levels are lower, but it doesn't stop entirely.

This causes a special problem for conifers. When the soil is frozen, roots can't get water and yet the upper leaves are transpiring. One way they cope with this is that the leaves are very small compared to deciduous leaves making them much more efficient in maintaining water levels. They are also coated in a heavy waxy material slowing down water loss even more.

One other factor affects water levels and that is wind. Many gardeners are concerned that wind lowers the temperature of plants, but that is mostly false. We feel colder in the wind because we are warm blooded and we call this the wind chill factor—a measure of how cold we feel. Plants are not warm blooded. They do not feel colder in wind. Wind does however increase the rate of water loss from leaves and branches.

Conifers are stressed in winter and if they can't maintain a high enough water level they abandon needles which usually show up in spring as brown needles. How does a gardener deal with this? They wrap trees.

Most people wrap their trees to keep them warm, which doesn't work. However, there is some value in protecting conifers from excess wind. To do this, don't wrap them. Instead, set up a windscreen six to twelve inches away from the plant and on the windy side. Reducing wind can prevent needle drop.

This might be a good idea the first year after planting because the tree does not yet have a fully developed root system, but it is not necessary in future years for hardy trees. Besides, what are you going to do in future years as the tree grows? Build bigger and bigger windscreens?

One thing does help a lot with water loss. Make sure the soil is well watered right up to the point where the ground freezes and water again in very early spring if needed. Mulch helps a lot.

9
Environmental Factors

Each type of plant has a preferred set of growing conditions. In nature plants tend to just grow where these conditions are suitable. If a seed lands in a spot with the wrong conditions, the plant may start to grow, but it is at a big disadvantage to other established species which usually results in the seedling being out-competed and dying.

Gardeners want to grow a wide selection of plants and they only have one location to grow them all in. Granted, even in a small garden, some areas are wetter and others drier, some have more sun than others, but for the most part each of us has one set of growing conditions and our selected plants must adapt to them.

This chapter will take a closer look at how plants adapt to various environmental growing conditions. Understanding that will help you make better plant selections.

Garden Hardiness Zones

Hardiness zones, also known as gardening zones, are a set of numeric designations that identify a narrow set of climatic conditions. Most of these systems are based on the average annual lowest night temperature. So for example, a hardiness zone 5 has a lowest night temperature in the range of –20°F (–29°C).

The original system was developed by the United States Department of Agriculture (USDA) which uses averages of the lowest night temperatures. Each zone represents a ten-degree Fahrenheit increment. In North America, there are a total of thirteen zones, which

Average Annual Extreme Minimum Temperature 1976-2005

Temp (F)	Zone	Temp (C)
-60 to -50	1	-51.1 to -45.6
-50 to -40	2	-45.6 to -40
-40 to -30	3	-40 to -34.4
-30 to -20	4	-34.4 to -28.9
-20 to -10	5	-28.9 to -23.3
-10 to 0	6	-23.3 to -17.8
0 to 10	7	-17.8 to -12.2
10 to 20	8	-12.2 to -6.7
20 to 30	9	-6.7 to -1.1
30 to 40	10	-1.1 to 4.4
40 to 50	11	4.4 to 10
50 to 60	12	10 to 15.6
60 to 70	13	15.6 to 21.1

USA Hardiness Zones.

are further divided into five-degree increments using the letters "a" and "b." For example, there's a Zone 10a and a Zone 10b, with Zone 10a being five degrees cooler than Zone 10b. Generally, nurseries don't bother labelling plants with "a" or "b."

The Canadian hardiness zone is based on low temperature as well as other growing conditions. It uses numbers similar to the USA system but there is not always a direct correlation between the two systems. You could live in US zone 4 and Canadian zone 5.

Europe follows along the same lines as the USDA Plant Hardiness Zones. The UK has a range from 7 to 9 but they tend to refer to two types of plants—hardy and not hardy. Since almost everything grows in zones 7 and 8, most plants are hardy. The Royal Horticultural Society also has a system of H-numbers. Zone maps are available for many global areas.

Each plant is also assigned a hardiness zone range to match the conditions where the plant will grow. In theory a zone 5 plant will survive winter in a zone 5 region but a zone 6 plant is likely to die.

Many plants also do poorly in zones that are too warm because they're not equipped to survive high temperatures or different levels of precipitation. Gardeners should select plants that include their zone within the range on the label or catalogue.

Much of online plant hardiness information is based on the US system, so it is useful to know your US hardiness zone, even if your country has its own system. If you are purchasing local plants in countries other than the US, make sure you know which system they use for labeling plants.

Limitations of Hardiness Zones

Cold hardiness is not the only important criteria. *Hydrangea macrophylla* has a hardiness zone of 4–9 and it certainly grows in my zone 5 garden. However, this plant blooms on last year's buds and they are only hardy to zone 6. So although the plant grows here, it rarely flowers because the buds get killed off in winter.

Creeping phlox (*Phlox stolonifera*) is hardy in zones 4–8. It is advertised as being an evergreen groundcover but it is really only evergreen in zone 6 and higher.

The blue poppy (*Meconopsis*) is hardy in zones 3–9 but I can't grow it in my zone 5 garden. I don't really know why because it does grow in colder regions not far from me. I suspect we are too humid in summer and don't have the reliable snow cover found in colder regions.

Hardiness of woody plants is also affected by something called provenance, which indicates where seed was collected. Redbuds grown from local seed are quite hardy here, but redbuds grown from seed collected in zone 8 will produce plants that are not hardy here. Provenance is rarely indicated on plant tags but many plants sold in colder climates are produced in warmer climates.

Hardiness zones are very useful for selecting plants, but they have limitations.

Establishing Plant Hardiness Zones

How does a plant get its hardiness zone? If you Google the internet you soon realize that commercial sites don't agree on a plant's hardiness. Values can be all over the map. The one site I trust the most is Dave's Garden (https://davesgarden.com).

When a new plant is brought to market, it does not have a hardiness zone designation. The breeder can certainly guess based on other similar plants. A new tulip cultivar probably has the same hardiness as other tulips. But what about a new hybrid where the parents have quite different zones?

One way that breeders establish its hardiness zone is to grow it in various climatic conditions, and the All-American Trial Gardens is one way to do this. These gardens are located in various zones in North America and test new plants. After a few years we start gaining some understanding about a plant's specific hardiness and over time the value gets better and better.

Gardeners are also important for fine-tuning hardiness zones. I tried growing the South African thistle (*Berkheya purpurea*) from seed even though it was listed as a zone 7 plant. It seems quite hardy in zone 5. Until gardeners try growing some of these warmer plants in colder conditions we don't know if they grow.

Dealing with Cold

Cold temperatures are a major limitation for plant growth especially below freezing, but plants have learned to adapt. Some plants have no problem growing in the Arctic.

So, how do plants cope with the cold? The answer depends on whether the plant is herbaceous or woody.

Herbaceous plants are lucky because their stems and leaves die back when conditions become unfavorable. Their underground structures are protected from cold by the soil and by the dry vegetation they drop. This vegetation traps snow around the crown, keeping plants warmer. It's one reason gardens should not be cleaned up in fall.

Woody plants go through a hardening off process where the soft green growth becomes hard woody material which is much less affected by cold. Deciduous woody plants lose their tender leaves.

Soil Is a Heat Source

Step outside in winter and inspect your poor garden. How can those perennials survive when everything is frozen? The secret lies in the center of the earth. The heat from hot molten lava travels up to the surface of the soil. The soil itself is a heat source.

If you touch the surface of the soil it is frozen and feels cold. Soil loses heat quickly to air and wind. However, the soil is quite a bit warmer even a couple of inches below the surface. Soil does freeze but it does not get much colder than the freezing point. The roots and other plant structures are much warmer than you think.

This heating effect of soil is dramatically increased when it's covered with mulch, fall leaves or better still, snow. New fluffy snow is about 95 percent air, making it as good an insulator as Styrofoam or fiberglass. Snow traps heat under it, keeping the soil warm. It also prevents wind from blowing heat off the surface of the soil.

Many plants will survive winter under snow but die without it.

Preventing Frost Damage

Pure water freezes at 32°F (0°C). As ice forms, its crystals are sharp enough to rip cells apart. The damaging effects of ice crystals harm plants more than the cold. So how do plants make it through a cold winter?

To understand this you first have to know a bit more about freezing water. We say that water freezes at 32°F but that is only true for pure water. Water that contains solutes (other chemicals) will freeze at a lower temperature. Beer is still liquid below zero because of the alcohol. The water in your car doesn't freeze in winter because it contains antifreeze. Salt is put on sidewalks in winter for the same reason: it dissolves in water, lowering the freezing point.

The sugars, proteins, fats and minerals in plant cells all act like antifreeze and lower the temperature at which the cellular liquid freezes.

All of these liquids will finally freeze at some temperature. This freezing point depends on the concentration of solutes in the water. A 3% beer freezes sooner than a 6% beer. Plants take advantage of this phenomenon by reducing the amount of water in their cells as winter approaches. Less water means a higher concentration of solutes and a lower freezing point.

Plants also use another trick called super-cooling, which prevents ice crystals from forming even when temperatures are below the freezing point. When all of these tricks are combined they let plants experience a temperature of about −40°F (−40°C) without freezing, and that is plenty to keep most plants alive.

A few woody species experience temperatures below this and still survive winter. They use an additional technique where they move water out of cells into the spaces between cells. The water does form ice there, but it does not damage the living cells.

Protecting Plants from Cold

There are things gardeners can do to protect plants from the cold. But there are also things gardeners do that do not actually work.

The most important thing a gardener can do is to buy hardy plants. Every fall I see questions about the need to wrap and protect trees. My answer is simple. If you bought hardy plants there is no need to wrap them—they are hardy. If you bought the wrong tree there might be things you can do to keep it alive, but why do that to yourself and the tree?

Keep gardens well watered. Even after leaf drop, roots are still actively growing and need water.

Here are some other things gardeners do to protect plants in the cold. Your climate will dictate if any of these are required.

Wrapping Trees

People wrap trees to keep them warm. Evergreens in the garden are wrapped in burlap. Containerized plants on a patio or balcony are wrapped in bubble-packing plastic. These things do not protect from cold.

Cold is the absence of heat, which is the energy stored in moving molecules. The only way to warm up something that is cold is to provide heat. If you take an ice cube and wrap it in a bubble pack, it is still cold because there is no heat source. It won't melt. In order for you to warm up the ice cube you have to add heat.

You can add heat in a few ways. If the sun shines on the clear plastic it will warm the inside and the ice will melt. If this is a container sitting on the ground you can get heat from the soil, but to make this work, the bubble wrap needs to go right to the ground. If you only wrap the top part of a container there is no heat source and the plastic does nothing to keep the plant warm.

You can also sink that container in the soil so that the pot is covered with soil. This is how I overwinter all of my potted plants.

What about wrapping an evergreen with burlap? Again there is no heat source so this does not keep the plant warm. It might reduce the wind, which reduces water evaporation, but burlap does not provide warmth.

You can easily check this yourself. Take a thermometer on a cold day and measure the temperature inside and outside the burlap. I've done it. Both are exactly the same because there is no heat source.

The other reason for wrapping evergreen trees is to prevent damage from snow load or ice buildup. Some vertical conifers like junipers are very prone to this kind of damage and the burlap keeps the branches close together and prevents snow or ice from bending them down. This technique does work but what do you do when the tree gets bigger? Are you going to use longer ladders and more wrap each year? A much better solution is to select trees that don't have this problem, like pines and spruce.

Mounding Soil

This used to be a very common practice for roses but it is used much less these days because we have very hardy roses available.

Soil is mounded around the plant in late fall and then removed in spring. In a semi-hardy rose you find that the woody parts and dormant buds under the soil will survive the winter and those above the soil die off. It is a bit of work, but is quite effective.

Rose Cones

Rose cones are Styrofoam boxes that are placed on top of plants like roses. The purpose is to keep the rose warmer in winter; it is the modern equivalent of mounding soil.

The Styrofoam won't prevent low temperatures, but it does prevent sudden drops in temperature and if the crack between the bottom of the cone and soil is sealed with soil, it does keep the plant warmer.

Do you need rose cones? They can help if you buy roses that are not hardy but it is better to buy hardy roses.

Mulch

Mulch adds an insulation layer above the soil and keeps below-ground structures warmer. It also slows down the evaporation of water from the soil and this may be of greater benefit to plants than the extra warmth.

Some argue that mulch freezes and then in spring keeps the soil cold longer than soil that is not mulched. This is quite true. But what is better, frozen soil or warmer soil? The answer is, it depends.

For landscape plants it is better to keep the ground frozen longer. You don't want it to warm up and get the plant growing, only to experience a late frost. It is better to keep plants dormant a bit longer. If they flower a week later it is no big deal.

Vegetable gardens are a bit different, especially in cold climates where the growing season is already too short. In this case it is better to remove the mulch as soon as possible in spring to allow the sun to warm up the soil. Many seeds and seedlings want warmer soil to grow.

I cover my garlic with straw in late fall to trap extra heat in the soil so roots keep growing as long as possible. They stay covered all winter. Then as early as possible, I remove the straw so the sun can warm the soil and get the shoot growing. I am trying to extend the growing season so they make larger cloves. Then in late spring when the soil is warm I put the straw back to stop weeds and to reduce watering needs.

Whitewashing Trunks

Winter sunscald happens in colder climates during the winter. The sun heats up the bark during the day, followed by a sudden drop to low temperatures at night. Rapid temperature changes cause damage to cells in the bark which results in the bark splitting. Most of the damage is found within a few feet of the ground.

Various wraps around the trunk, either paper or plastic as well as white paint, reflect sunlight and therefore keep the bark from getting too hot. Wraps will also hold in some warmth at night but this will have a very limited effect.

Brown-colored paper wraps actually absorb heat and can increase the temperature of the bark. Only white or silver products reduce temperature extremes.

Wraps should be removed in spring. Painted surfaces can be left alone.

Dealing with Heat

Plants do not acclimate to heat as readily as to other stresses like drought or cold. At most, a plant can tolerate a few degrees above the maximum temperature in its native habitat. High temperature increases evaporation and speeds up moisture loss which causes leaves to wilt and even drop in severe cases. Other forms of heat injury include lower photosynthesis and respiration rates, unstable cell membranes and damage to proteins and DNA.

High temperatures tend to speed up chemical reactions and damage enzymes. To prevent this, plants, animals and microorganisms produce proteins called heat shock proteins (HSPs). These act like a protective coat that stabilizes and protects enzymes and nucleic acids from degrading.

Soils, especially bare soils, become unusually hot under high temperatures. This can prevent seeds from germinating and slow down root growth.

Container plants are especially susceptible to heat. In recent years cloth bags have been promoted as a way to keep soil in containers cooler. Unlike plastic, the cloth breathes and water evaporates from the sides causing a cooling effect. In my own tests, the temperature of soil inside these is at most 1.8 degrees Fahrenheit cooler, not enough to make much of a difference.

Some plants like cactus and succulents are designed for hotter climates. An important adaptation is having more saturated fatty acids in their membrane lipids. Lipids are fats, and all cell membranes are made up of a double layer of lipids that hold water inside the cell. Saturated fats have a higher melting point than unsaturated fats (think butter versus olive oil), so cell membranes with more saturated fats maintain their structural integrity better at higher temperatures.

To acclimate to heat in the short term, plants will orient their leaves vertically or even roll them up to minimize sun exposure and prevent sun damage. Leaf rolling occurs primarily in grasses and reduces their surface area. Finely dissected leaves with many small lobes maximize heat loss compared to less-dissected leaves. Many plants grow hairs (trichomes) to shade the main part of the plant.

Pests and pathogens can be more active at higher temperatures. A higher metabolism in insects results in higher feeding rates and more rapid reproduction. Fungi and bacteria also reproduce faster at higher temperatures increasing infection rates. The plant may need to expend more energy defending itself from potential invaders and attackers.

Dealing with Water Extremes

There's such a thing as too much of a good thing. While water is essential for plants, too much water can be harmful and even deadly. Recall that plants, and beneficial soil microorganisms, need oxygen to produce energy through cellular respiration. Healthy soils contain about 25% air and 25% water which is perfect for plant growth.

After a heavy rain much of the air is pushed out of the soil and the air spaces are filled with water. This makes it hard for roots to get oxygen, but they can endure this condition as long as it does not last too long. As time passes, water continues to seep both down and sideways in the soil and plant roots take up water. Normally it does not take too long before larger spaces are again filled with air.

Things don't work quite so well in heavy clay soil and compacted soil. These types of soil already have few air spaces and they hold onto water longer. Some plants can't live in such conditions and their roots start to rot due to a lack of oxygen and soon the plant dies. Other plants are much more adapted to heavy wet soil and are able to cope.

At the opposite end of the spectrum, it doesn't rain for weeks or even months. These soils get drier and drier. First the large channels in the soil drain of water and fill with air. Soon even the small channels are empty. A thin layer of water still coats each soil particle and now roots have to work much harder to absorb this water.

A lack of water will cause plants to shut down unnecessary metabolism. Flowers and fruits are aborted or their development never starts. Growth then slows down and in later stages stops completely. Root hairs die off. Leaves start to drop. The plant is starting to go into an emergency dormant stage. If water does not return in time, the plant might die.

You see this in North American ephemerals like trilliums which grow best in moist cool conditions. If they get a steady supply of water, their leaves remain green and active much longer. On the other hand, if late spring is dry, they pack it in early and go underground. They are quite adapted to spending late summer and fall under ground.

If the summer is extra dry, many perennials will go brown and drop their leaves early. This does not harm them very much provided they had a long enough season to build up a good food reserve.

Some plants are much better adapted to dry conditions. Cacti and succulents can go many months without water. They simply shut down metabolism, which is low at the best of times, and wait it out. The plants do get thinner very slowly as they lose water. Eventually they even look wrinkled, but even at this stage they can quickly recover when rains return. Rain causes root hairs to grow quickly and they start absorbing water. It is just another normal drought in the desert.

Many flowering bulbs are very dry during the summer in their native habitats so they have adapted to growing and blooming in spring or fall and spending the summer dormant underground. One of the most unusual of these is the clochicum, commonly, but incorrectly, called the fall crocus, which makes leaves in spring and flowers in fall.

Watering

Watering seems so simple but it is one of the hardest things for new gardeners to learn.

Many people water on a schedule. They water the houseplant every Thursday. They water the perennial bed every Friday. They're following common advice in books and online that suggests watering weekly. That makes no sense.

From the above discussion it is very clear that you should water once the soil reaches a certain state of dryness. The timing of this varies greatly depending on many factors including the type of soil, the temperature, the humidity, the size of the root ball, the size of the plant and the amount of light the plant gets. Not only are these conditions different in different gardens, they also change with the seasons.

Don't water on a schedule. Water when the soil starts to dry. How do you know when this is? Stick your finger in the soil. If it feels wet, don't water. If it feels dry, water. This will suit 98% of the plants you grow.

There is also some variation in the plants and you will understand this better as you gain experience. Orchids, cactus and most succulents can go completely dry with no ill effects. In winter cactus are left dry for a month or more. Some houseplants want to stay wet all the time and get droopy when the soil is even slightly dry. African violets and streptocarpus don't want to be too wet or they start to rot, but sitting dry for a day or two does no harm.

The most extreme examples of drought-tolerant plants are called resurrection plants, which survive without moisture for months or even years. One of the most famous examples is *Selaginella lepidophylla*, commonly called the rose of Jericho or simply resurrection plant. This plant can lose up to 95% of its moisture content without damaging its cells and it can remain dry and alive for several years. During drought periods, the plant goes dormant and curls inwards to protect its vulnerable reproductive structures from desiccation. The plant will then produce a special crystallized sugar called trehalose, which keeps the salts within the plant from damaging cells as the water content decreases. When the plant receives water again, the trehalose crystals dissolve and the plant's normal metabolism reactivates. The plant will then uncurl, reveal green leaves and release seeds.

Drought Affects Roots
Drought affects how roots grow. In drier conditions roots of many plants grow longer with less branching and they tend to grow deeper in the soil. This can be affected by the watering practices of gardeners. Watering less but more frequently produces shallow roots because that is where they find water. Watering deeper and less often produces deep roots. Either type of root grows good plants, provided watering is maintained. If watering is suddenly stopped and the weather is dry, the deeper roots provide water longer for the plant.

A good example of this is your lawn. Frequent short watering periods result in shallowly rooted grass that is not very drought tolerant. Deeper, less frequent watering is better. In cool climates no watering is an even better option because cool-growing lawn grasses are well adapted to going dormant in summer.

Adaptability of Plants

Garden success depends a lot on a plant's ability to grow in its new environment. There are two terms used to describe how a plant modifies itself to grow in a new location: adaptation and acclimation.

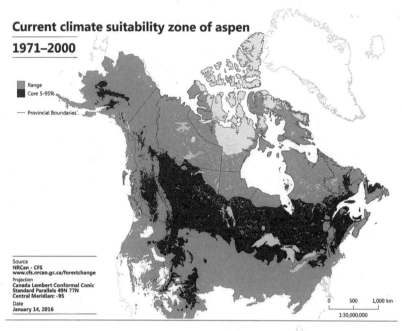
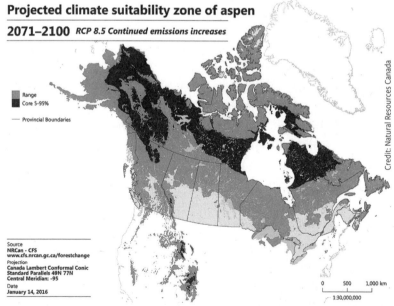

Expected migration of aspen trees due to climate change.

If someone says a plant is adapted to its climate, or has adaptations suited to its environment, that means the plant has changed genetically to allow it to grow better in a new environment. These changes take time and occur over many generations.

When plants modify their growth habit for a new environment in your garden it is called acclimation. This is a process whereby plants undergo physiological changes to maximize their growth potential. Acclimation can happen quickly in a matter of days or months.

Plant breeders do influence the adaptability of plants by breeding them for a number of generations or by introducing new genetics using new species. Rather than waiting for nature to select characteristics they speed up the selection process. For example, they might select shorter, stockier plants that have stiffer stems to be more wind resistant, so they don't need staking. They also select for better cold tolerance so that the plants can be sold over a larger climatic region.

The combination of adaptations and short-term acclimations allows plants to occupy a wider range of environments. In nature, many plant species inhabit a broad geographical range, but the species will grow and spread more vigorously in their ideal conditions. For example, the trembling aspen (*Populus tremuloides*) has a native range that extends across Canada and into Alaska. It's well adapted to cold climates but it doesn't grow as well in milder climates like southern British Columbia. Their southern range in the east is southern Ontario and a few northern states. In the southern range, trembling aspens don't live as long as farther north and their wood is weaker, making them a poor choice for landscaping.

Finding a more suitable location is another way plants adapt to changing environments. Though individual plants can't move, they can technically migrate to more suitable conditions. This is mostly a function of chance. A seed lands somewhere and if conditions are suitable a new plant grows. If the new location is not ideal or if competition is strong the new seedling will not do well. On the other hand the new location might be more suitable and it thrives producing even more seedlings in the new location.

This migration of plant species happens all of the time, but climate change is accelerating the process. In the case of the trembling aspen it is predicted that warming conditions will result in the majority of trees in their current locations to completely die out by 2100, and that they will be growing in a region that is north of their current habitat.

Plant adaptations and acclimation are incredibly important to our gardens. Gardeners can grow plants from different countries that have adapted to a similar climate and can reasonably acclimate to novel conditions. Adaptations can be bred into garden plants to improve their chances of success in novel environments, or to resist damage from stress, pests or disease. But there's a limit to how much a plant can adjust to their environments.

How Climate Change Affects Gardens

By now, nearly everyone has heard of climate change and its consequences. Global warming is often equated with climate change, but climate change goes beyond increased average temperatures and includes the increased frequency of abnormal precipitation and wind patterns. The term "weather" is often used interchangeably with climate, but weather refers to the short-term state of the atmosphere and climate is a region's long-term average weather conditions.

Climate change is already affecting everyone in subtle (and not-so-subtle) ways and gardeners are no exception. Plants rely on the environment for signals relating to the timing of growth, flowering, reproduction and dormancy. Higher average temperatures can confuse plants—spring-blooming shrubs bloom earlier and wilt sooner in the season, spring bulbs could finish blooming and die back sooner than expected, and summer-blooming plants might start blooming too early.

Of course, climate change results in abnormal weather, not just warmer temperatures. This means that spring weather can be abnormally cold, dry or rainy or shift between cold and warm and wet and dry. These sudden shifts stress plants out when they've just begun growing and are at their most fragile. Plants can become so stressed they wilt, slow their growth or die back. Shrubs and trees may break bud dormancy too early, increasing the risk of damage from spring frost. There are a few upsides to warmer and cooler weather though: spring plants bloom for longer in cool weather and summer and autumn plants can continue growing and blooming much later in the season.

Gardening relies heavily on timing, and gardeners are always keen to create a succession of flowers for multi-season interest. Scientists compared today's flowering times for woodland plants near Concord, Massachusetts, to the times recorded in Henry David Thoreau's famous book *Walden* and found that plants are now flowering about 18 days sooner than in the mid-1800s.

It's a bit early to measure the cultural impacts of climate change, but it's possible that various regions could eventually lose iconic native plants. Imagine if southern Ontario becomes too hot and dry for trillium! All of these changes may eventually influence where plants can be grown, altering plant selection. Plants may get reassigned to be grown in different regions by horticulturists, botanists and nurseries. It might be possible to eventually grow plants from much warmer regions in what are now colder climates, or for the already established gardening zones to change as extreme regional temperatures shift.

Even in the garden, plants are part of a wider natural system. Climate change is a threat to our favorite plants because of the potential impacts to local ecological processes. For instance, invasive species could colonize new areas or become even more aggressive if the conditions shift to become favourable to them. A good example of this is the popular miscanthus grass. Most cultivars do not set seed in zone 5 because the summers are too short and so it is not an invasive problem here. However in zone 6 and warmer it is starting to become a real problem.

Higher levels of carbon dioxide in the atmosphere can boost plant growth, but aggressive and weedy plants usually exploit excess resources better than desirable garden plants. Insects, including pests, typically reproduce more rapidly and successfully in warmer conditions. If winters become shorter and milder, more invasive plant and insect species could thrive.

Non-plant species also have relationships with plants, especially insects, birds and mammals that are key to plant reproduction. Changes in temperature and moisture can alter flowering times, causing pollinators to arrive too early or too late to fertilize the blossoms. Some species are so specialized that they rely on a specific plant to complete their life cycle. The symbiotic relationships between plants, insects and animals are imperative to the success of gardens and the overall health of ecosystems. Climate change poses a direct threat to these relationships.

What Can Gardeners Do about Climate Change?

Aside from limiting our ecological footprints at home and at work, gardeners can take specific approaches to make their gardens more resilient to climate change as well as more climate-friendly overall.

Plants, especially trees and shrubs, are key to mitigating climate change as they reduce the CO_2 in the air. Larger plants take up more CO_2 than small plants. The more leafy plant material a garden produces the better it performs for reducing CO_2 levels. Growing lots of vigorous plants rather than a sea of turf grass is already an improvement.

Common gardening habits like tilling the soil are bad for soil health and for the environment. Tilling not only destroys important soil structure but it adds excess oxygen to the soil. This increases microbial activity which uses up the stored carbon and releases CO_2. This is a major problem with modern-day agriculture but many farmers are moving to no-till or low-till scenarios. There is no reason for gardeners to till their soil, except maybe in year one when making a new garden and even that can be done in a better way using sheet mulching.

Gardeners have a habit of fertilizing gardens that don't need to be fertilized and I am talking about both synthetic fertilizer and organic fertilizer like manure and compost. Use of these materials contributes to climate change. Unless you have a known nutrient deficiency, you probably do not need to fertilize. Vegetable gardens and containers are a bit different. I grow about 3,000 different types of plants and don't fertilize any of them. I do add a bit of nitrogen to the vegetable garden because I have a short growing season and want faster growth and I fertilize containers because they get watered a lot.

Hardiness zones have changed in some areas but it is important to understand that these zones have a 10 degree Fahrenheit range. Since 1900 global warming has increased by about 1.8°F (1°C). Although this change in climate is significant, it is not very large compared to hardiness zones. In most regions you will not see significant changes in the plants you can grow.

The other thing to consider is that hardiness zones are based on average annual lows. If climate change results in more extremes, you might actually experience more extreme lows, which negates any general warming trends. Don't start planting less hardy plants.

So, what is a gardener to do? The answer lies in diversity, which is how nature copes with environmental change. Planting different species and varieties of plants reduces the extent of damage to your whole garden from sudden and extreme shifts in weather or new pests and diseases. More diverse plants also support a wider range of beneficial insects and pollinators that will keep your garden blooming for years to come.

Choosing plants that are more flexible to different conditions can make gardens more resilient. Some examples are riparian plants that have adapted to alternating periods of flooding and dryness, drought-tolerant plants from grasslands and savannahs, high-elevation plants or even plants found in harsh environments like screes and alvars. Some garden plants tolerate a wider range of conditions compared to others and may be safer to plant than ones that prefer a very narrow range of conditions.

Many people promote native plants and they can be a good choice. But it is important to understand that native plants cover a wide spectrum of requirements. Some are easy to grow, others are very difficult to accommodate. Natives can also be aggressive spreaders that you really don't want in a garden. Pick natives based on plant qualities, not because they are native. And remember climate change will also make it harder for some natives to remain native where you live. My aspens are slowly dying out.

10
Growing from Seeds

Seeds play a minor role for most beginners except maybe in the vegetable garden, but as gardeners gain more experience they realize that seeds open up access to a vast range of plants at an affordable price.

I have grown several thousand different species of plants from seed and two-thirds of those are not available from nurseries. The greatest joy I get in gardening is watching the first flower open on a species I have never seen before.

This chapter is split into two parts. The first part will help you understand seeds and the second part will show you the best ways to germinate them.

When Is Seed Mature?

Maturity can be defined in different ways. One way to define it is to consider maturity as being the point where the seed can germinate. To appreciate this better it is important to understand orthodox and recalcitrant seeds.

Orthodox vs. Recalcitrant Seeds

Seeds have been classified into two general groups: orthodox and recalcitrant (non-orthodox).

The first group, orthodox seeds, probably got their name because these seeds behave very much like the seeds that have been collected and stored for thousands of years. After maturing they can be dried and stored for a long time. This group makes up 80% of all seeds.

The second group of seeds were researched more recently and became known as recalcitrant (having an obstinately uncooperative attitude). In the early days these seeds seemed impossible to germinate, but now we know that they die when they dry out or are stored too cold.

This sounds pretty clear—seeds are either orthodox or recalcitrant—but it is not that simple. We now know that there is actually a third category, sub-orthodox, which is halfway between these two extremes. There are also exceptions in each category.

Mature Seed

Almost everything you read about gardening will tell you that seed is mature when the seed pod gets brown and dry. The concept is easy to understand and works well for the general public but seed maturation is more complex than this. Many orthodox seeds continue their maturing process after the seed is black and/or released from the fruit.

Seed from chili peppers (*Capsicum annuum*) have the highest rate of germination when they are left in the fruit for another 14 days after harvest. This is true for green, yellow and red fruit, with higher germination from ripe red fruit. This clearly shows that for this seed, maturation was still taking place after harvest. Similar results were found for tomatoes, eggplants, watermelon, bell peppers and cucumbers.

Orthodox seed can also germinate before they look mature. Soybean and corn can germinate 20 and 50 days respectively prior to full maturity, but the resulting seedlings are smaller and weaker than seedlings grown from more mature seed.

When you place freshly collected seed in the fridge, some of the chemical processes that are still taking place inside the seed suddenly stop. The maturation process is halted, and you have just stored seed that is not fully ripe. For this reason, it is not a good idea to store your seed in the fridge immediately after collecting it. Let it sit at room temperature and continue its maturation process. By late winter or spring it will be ready to germinate.

Recalcitrant seeds need to be treated differently. These seeds die if the moisture level gets below 30%. Because of this high moisture content they can't be frozen. Most of these seeds can be stored at 32°F (0°C), but some tropical seeds are damaged even below 59°F (15°C).

Plastic bags containing a bit of moistened vermiculite work well to keep the moisture level up. Even with this type of storage, these seeds tend to have a short life span of a few weeks to a couple of years. Even seed banks don't have an easy way to store them long term.

Recalcitrant seeds tend to reach germinability sooner than orthodox seeds.

When the Norway maple (*Acer platanoides*), an orthodox seed, is compared to the sycamore (*A. pseudoplatanus*), a recalcitrant, researchers found that the sycamore reached germinability ten weeks before physiological maturity (seeds look ripe). The Norway maple reached germinability four weeks before physiological maturity.

A recalcitrant seed, *Galanthus nivalis*, the common snowdrop, has some germinability 29 days before natural seed dispersal, with a maximum germinability of 79%, 12 days before dispersal.

Many recalcitrant seeds germinate better when the seed is collected in the green stage, rather than the normal brown/black stage. By the time these seeds look mature, they are already getting old and losing viability.

The Seed Germination Process

Once seeds are mature, we consider them to be in a state of dormancy. Think of dry bean seeds that can sit in a jar for years and they'll still grow when you plant them. Some 2,000-year-old date palm seeds were found at the edge of the Dead Sea and they germinated. A perennial silene species was grown from a 32,000-year-old seed found in a squirrel's nest in Siberia. In the right conditions, certain seeds can remain dormant and viable for a very long period of time.

It is important to realize that seed is a living organism. It is taking in oxygen, metabolizing sugars and producing carbon dioxide. The key to its survival is that the low water level keeps the seed in a type of hibernation, not unlike that of a bear in winter. The metabolism is slowed way down, but it has not stopped completely. Even that 32,000-year-old seed is still biologically active.

The seed will remain dormant until conditions are right for it to germinate, a process called breaking dormancy. Once this happens the seed embryo starts growing a radicle (a small root) which emerges first, followed by a shoot.

The seed is able to grow a small root system, a stem and some initial cotyledon leaves without any nutrients or light from the outside world. All of the food and energy needed for this growth comes from the food stored inside the seed, resulting in a new growth that is many times the size of the seed. The idea that a complete plant can be grown from a small hard nugget of cells has to be one of nature's best creations.

The Mysterious Cotyledons

The cotyledon is the first leaf a seedling will make. They are actually formed inside the seed and are not true leaves, even though they look and function somewhat like leaves. They are also called the embryonic leaf or seed leaf.

About 75% of angiosperms (flowering plants) produce two cotyledons and are called dicotyledonous or dicots for short. The other 25% produce only one cotyledon and are called monocotyledonous, or monocots (grasses, orchids, palms, iris, lilies).

The gymnosperms (non-flowering plants) have a varying number of cotyledons. For example, a Douglas fir has seven.

A germinating seedling may or may not push the cotyledon above the soil level. If they are above the soil line it is called epigeal germination and if below the soil line it is hypogeal.

Tomato seedling with two cotyledon leaves and two true leaves.

Why does all this matter? Knowing the above will help you figure out if your seedlings are growing correctly. The first thing to note is that the cotyledons do not look like true leaves, so if your new seedling does not have the leaves you expect, don't worry until the next set of leaves emerge. Many seedlings have cotyledons that look alike.

Onions and most bulbs produce a single grass-like blade when they germinate. This is the single cotyledon of a monocot. The true leaf will develop later from the base of the seedling.

In many cases the cotyledon gets trapped inside the seed coat and gardeners worry that the plant won't grow correctly and try to remove the seed coat. That can cause a lot of damage to the seedling and even kill it. Just leave the seed coat alone until nature takes it off.

In some cases, like peony seedlings, the cotyledon never shows outside of the seed coat.

The cotyledon may photosynthesize and start making food for the plant, but its role is mostly to provide stored food for the initial growth of the seedling.

Once you see the start of the first true leaf forming, it is time to provide the seedling with both light and nutrients. The stored food in the seed has now been used up and the seedling needs to produce its own food.

Why Do Seeds Stay Dormant?

Most plants in temperate regions form mature seeds in late summer or fall. If the seed germinated right away, many would not survive winter. It is much better for the plant to wait until spring before it germinates. This is one of the main reasons plants have developed a number of tricks to keep the seed dormant until spring.

Deserts have a different problem. They might have cold winters, but the real challenge is access to water. The seeds of many desert plants go dormant until they get a substantial rain event. Then the seeds quickly absorb water and germinate.

What happens if all the seeds germinate at the same time and then there is a drastic weather event that kills all of the seedlings? This could wipe out a species in a local area. Plants overcome this potential problem by producing seed that stays dormant for various time periods, ensuring they don't all germinate at once.

A New Way to Look at Dormancy

Consider dormancy from a gardener's perspective. A gardener looks at seeds from the outside. We don't cut seeds open and dissect them to identify small internal changes that may be taking place. So the following may not be 100% botanically correct, but I think it makes it easier for gardeners to understand dormancy.

I see seed development and dormancy as part of the plant development process. The life of an animal has several distinct phases: fertilization, embryo development, birth, childhood and adulthood. This is not unlike the phases of plant development which consist of

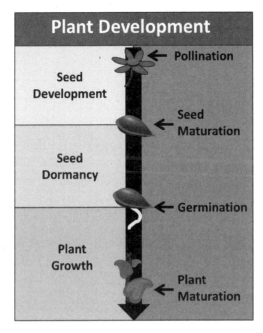

Stages of plant development.

fertilization, embryo development, seed maturation, germination, seedling stage, adult plant.

The plant's equivalent stage to an animal birth is the seed maturation point: the point in time when the seed separates from the mother plant. It is now a separate entity with a life of its own. It now enters a phase we call seed dormancy, but is in fact clearly a phase of plant development. The root and shoot are undergoing chemical and physiological changes—we just can't see them because they occur inside the seed coat.

Germination is simply the point at which we humans are able to see something happen outside the seed; it is not the birth of the plant.

Seed dormancy is the period of time between seed maturation and germination. Many chemical and physiological processes take place during this time period but there is no period of inactivity where nothing is happening.

It is also important to understand that the complete process from fertilization to plant maturation is a continually changing process. At no point is the plant metabolically dormant.

Breaking Dormancy

Every seed seems to require its own process for breaking dormancy. Some need to be stratified, some scarified and some need cold-warm cycles.

In almost every case the first step in breaking dormancy is the absorption of water. This makes sense since almost all chemical re-

> ### Plant Myth: Seeds Can Have Double Dormancy
>
> Double dormancy is defined as a condition where seeds need to overcome two or more primary dormancies in order to germinate.
>
> If you plant some peony or trillium seed in the fall you won't see any green growth until the second spring—if you're lucky. This is routinely described as an example of double dormancy—the seeds need two cold periods before they germinate.
>
> Peony seed will not germinate immediately because collected seed has an embryo that is not fully developed. It requires a few months in warm conditions to finish developing and then roots emerge. Once roots develop they need to reach a certain size and then experience a cold period before the shoot starts to grow.
>
> Do peonies show double dormancy? The initial warm period is a dormancy that needs to be overcome. After that the seed germinates which is a single dormancy. Germination is defined as the point where roots appear. It is not defined in terms of when something shows above ground.
>
> What about the dormant shoot? This is definitely a dormancy, but it is a dormancy of plant growth and takes place after germination. Adult peonies have similar annual dormancy periods.
>
> Many of the plants that are claimed to be double dormant are not, especially if treated correctly.

actions in a cell occur in water and the seed is now very dry. Water causes cells to expand and you can easily see this if you soak some dried wrinkled peas overnight. They become smooth and twice as big.

As cells expand, all kinds of chemical reactions are set in motion. Protein enzymes are created and hormones are launched into action. The seed starts to wake up but it might still need other triggers to start the germination process.

Germination Inhibitors

During seed development, chemicals are produced that prevent germination. These inhibitors need to decay or be converted by enzymes so that germination can take place.

Many fruits contain these inhibitors and as long as the seed is inside the fruit, they won't germinate. When birds eat these fruits the stomach acids clean the seeds, removing the inhibitors, and the deposited seeds germinate more easily.

Many gardeners will rub the gel off tomato seeds when they collect them for the same reason. Some will let tomato mush sit for a few days to ferment in an effort to remove germination inhibitors. I did some trials to see if this is a requirement, and it's not. Seed with dried gel on it germinates just as well as cleaned seed.

Light Effects

Some seed requires the presence of light, or its absence, to germinate. Light-sensitive proteins are either activated or deactivated, depending on the seed. These proteins cause other reactions that initiate germination.

Most light-sensitive seeds need light, but a few need a dark period to germinate.

The need for light makes a lot of sense for plants. If the seed gets buried too deep in soil, there is no point in having it germinate because the new shoot won't reach the surface of the soil to get light, and the seedling will die. If a kind gardener, hoeing the soil, moves the seed to the surface the light triggers germination.

The amount of light needed for germination is very small and regular room light is usually plenty.

Hormone Balances

Seed dormancy appears to be very dependent on the ABA/GA3 ratio, two hormones called abscisic acid and gibberellic acid. These ratios change over time until they reach a point where germination can happen.

These hormonal changes occur naturally and are a kind of germination delay mechanism allowing seeds to germinate over several years instead of all at once. Gardeners can speed up the process by adding GA3 manually.

Chemical Absorption

Water, nutrients, hormones and other chemicals can be absorbed by the seed. For example nitrate salts, peroxide and gibberellic acid have been used to shorten the dormancy period.

Changes in Seed Coat

Some seed coats have a waxy outer covering that slowly decomposes. Seeds can also have physical and chemical changes take place inside the seed coat.

Some seeds have a very hard seed coat that needs to be physically damaged before water can be absorbed. Walnuts would be an example of this, but even small seeds can have very hard seed coats. Gardeners can manually damage the seed coat by using a technique called scarification. Sandpaper and nail clippers work well for this.

Temperature Fluctuations

This is one of the most common requirements for plants native to temperate regions. They need some form of cold-warm-cold cycling.

Think of the seed that matures in late summer. It falls to the ground and experiences a warm fall temperature and maybe even fall rains. If the plant germinates now, it might be killed in winter so the seed is pre-programmed to require a cold period (winter). The seed remains dormant until temperatures warm up after the cold spell. That way it germinates in spring and has all summer to grow to get ready for winter.

Other species do it differently. They take advantage of the warm fall and germinate right away. They then remain hidden underground until spring.

The white trillium uses a hybrid of methods. In any given seed pod there are some seeds that germinate in fall and grow only a tuber underground. It does not waste its resources on making leaves that will be damaged by winter cold. These tubers then need a cold period before they grow their first leaf, next spring.

Other seeds in the same seed pod are programmed to require one or more cold periods before they even germinate. Some will germinate the first spring after being formed, and others will wait two years. Different seeds within the same seed pod have different requirements to break dormancy. This is a good strategy for species survival, but makes it more difficult for gardeners.

Stratification is the process of providing artificial temperature cycles and is commonly used by gardeners to germinate seed. The tricky part is to know which type of cycle works best for a given seed. One of the best sources for this kind of information is the Germination Guide found on the Ontario Rock Garden and Hardy Plant Society website.

Seed Storage

Common garden advice is to store seeds in the fridge. Some even say that you should freeze them. There is no "best" method for storing seeds. The method you use depends on the type of seed you are storing as well as how long you want to store them.

At a minimum you first need to determine which type of seed you are dealing with: orthodox or recalcitrant.

How do you determine this? Unfortunately, that is not easy. I have not found any good lists, but Bill Cullina, of the New England Wild Flower Society, has some plants listed; he uses the term "hydrophilic" instead of recalcitrant. BotanyCa also does a good job of identifying seeds that need moist packaging and Genesis Nursery, Inc. also has a list. If you can't find your plant on a list, here are some general rules you can follow:

- 80% of all seeds are orthodox.
- Most vegetables are orthodox.
- Many North American woodland plants are recalcitrant.
- Some willows, poplars, elms, maples, oaks, hazels, walnuts, chestnuts and hickories are recalcitrant.
- Many tropical rainforest plants are recalcitrant.

If it is an orthodox seed and it will be germinated within the next year, store it in paper at room temperature. If your goal is to store the seeds for many years, make sure they are very dry and have had enough time to mature (several months), and then store them in the fridge—not in the freezer.

Recalcitrant seeds should be stored moist. Use a temperature that is similar to their native environment. For temperate seeds, store at outdoor temperatures, and gradually cool them down as winter approaches. Then store in the fridge. For tropical seeds, store between 59°F (15°C) and 68°F (20°C). In both cases it is best to plant as soon as possible for best germination. Consider collecting some seeds in the green stage to see if germination improves.

There is huge variation in seeds and their growth behavior. It is always a good idea to research each seed and follow the advice given for it.

Best Method for Starting Seeds Indoors

There are several different methods for starting seeds indoors that work well for vegetable seeds and flower seeds. None of the seed-starting methods are right or wrong—they all work and produce results. The best method is the one that aligns to the right type of seed and your home environment.

There are three basic ways for homeowners to start seeds: in pots or containers, the paper towel or baggy method, and winter sowing. Each method has a number of variations.

The descriptions in the book are brief due to limited space, but a list of detailed videos are provided at the end of this chapter.

Starting Seeds in Pots and Containers

Using this method you take a pot or container of some type and fill it with soilless mix. Plant the seeds in the pot, water it and wait until they germinate. This is the traditional method for starting seeds, and it works very well for all types of seeds including vegetables, flowers and trees.

Pros:
- It is a simple method. You can use plant pots or even recycled containers as long as you punch some holes in the bottom. You can use garden soil but buying a seedling mix is a better option.
- Once the seed germinates it can grow right in the same pot until it is ready for the garden.
- Works well for small and large quantities of seed.
- A good method for indoor plants and garden plants.

Cons:
- You can't see the seed germinate. If nothing grows, it is difficult to figure out why. Was there no germination—the seed might have been dead? Did the seed germinate but then die—maybe you have a fungus or pH problem?
- Cold stratification is difficult in pots unless you have a dedicated fridge for plants or a cold fruit cellar.
- Seeds that take a long time to germinate need to stay in the pot a long time. This becomes a lot of work in summer when the pots dry out quickly and need to be watered every day.
- A light source is needed to grow the seedlings. Windows provide a very low level of light compared to the sun. New LED lights provide much better light. Growing under poor light results in etiolated (tall and weak) seedlings.

This method can be modified by using Jiffy pellets, which are special disks of peat moss that expand in water. A plastic mesh holds the peat into the form of a pot. These are inexpensive and popular among beginners, but they are difficult to use, are not nearly large enough for most seedlings, and the mesh needs to be removed before potting into another container. I don't recommend them.

Many online sources suggest using egg cartons, egg shells and even ice cream cones. These are all gimmicky and do a poor job. Get yourself some three to four-inch plastic pots. They will last for ten years or more and are large enough so they don't stunt the growth of seedlings. All of mine are repurposed from purchased plants.

Starting Seeds in Paper Towels or Baggies

The paper towel and baggy method are very similar. The traditional method takes a paper towel, places seeds on top, folds it in half and inserts it into a plastic sandwich baggy. A bit of water is added to make the towel moist.

This is my preferred method and I have developed an improved version of this method which can be seen in my video: https://youtu.be/dirzoWIMQio.

When the seeds germinate, remove them, pot them up and grow them normally.

Pros:
- You can see the germination process. Not only is this exciting, but it can tell you a lot about your seed. If you never see a developing root in the baggy you know that the seed is either not viable or the pre-treatment was not the right one. If it germinates and produces a root, then it is viable. If the seedling subsequently dies, it is not a germination problem.
- A lot of seed can be germinated in a small space using this method. You can hold the baggies for 100 different seeds in one hand—try doing that with 100 pots. Granted, if you are successful with all 100 seeds, they do need to go into pots at some point.
- Seed that takes a long time to germinate requires little care since the paper towel stays moist in the baggy for a month at room temperature.
- Stratification procedures are easy to carry out since the bags take up so little room in a fridge.

- Maximum use of seed. Since you can see which seeds germinate, you need fewer seeds. In the potted method most people plant excess seeds and weed out the extra. With this method you can put each seed into its own pot once it germinates. This can be a real benefit for rare or expensive seeds.

Cons:
- Requires an extra step. You have to put seeds into baggies, and then you still need to pot them up. But you only pot up the ones that germinate.
- Extremely small seeds can be difficult to handle but I just rinse them onto the soil once I see some germination.
- No special lights are needed for germination, but once seeds are potted up they do need the same light as any growing seedling.

Vermiculite and Baggies

This is a variation of the above baggy method. Instead of using a paper towel in the baggy, it is filled with vermiculite that has been barely moistened. You can also use peat moss or seed-starting soil instead of the vermiculite.

This works in much the same way as the paper towel—it keeps seeds moist.

Vermiculite has the advantage that growing roots and shoots do better in vermiculite than in the small air space between the paper towel and the plastic. This means you don't have to look at your seeds quite as often since they can grow for days or even weeks in the vermiculite, provided there is some light.

The down side to this method is that you can't see smaller seeds as easily. I find vermiculite works great for large seeds, but I don't like it as much for medium and smaller seeds because I can't see them.

Winter Sowing

The word "winter" in "winter sowing" refers to the fact that seeds are germinated and grown outside in winter. Seeds are planted in containers that will act like little greenhouses. After planting, the containers are set outside, and left until plants are ready to be transferred to the garden.

Pros:
- Minimum care. Once the seeds are planted and the containers are outside, there is virtually no maintenance of the containers. Later in spring you might need to water a bit.
- The sun is your light source.
- Seedlings are short and tough because they grow in cooler conditions with lots of light. They are much stronger than seedlings grown indoors.
- Stratification is not needed since the seeds go through a natural stratification process during the winter and spring periods.
- Gardeners routinely use garden soil, which is less expensive than purchased seedling mix. A clay soil works well.

Cons:
- The method is not suitable for seeds that are damaged by frost. It is not suitable for some tropical plants and most indoor plants.

Winter sowing.

- You can't see the seeds germinate. If nothing grows it is difficult to figure out why.
- Less suitable for seeds that need a warm cycle before a cold cycle. If you start early enough, the baggy method could be used to provide a warm cycle, followed by winter sowing.
- Anything that does not germinate by spring needs to be maintained and watered all summer, or discarded.
- There is no control over temperatures which makes it hard to start plants early in a warm environment to get a head start on the season. For example, it is better to start tomatoes inside where it is warm.

Which Seed Germination Method Is Best?

The best method is the one that meets your specific needs. Winter sowing is easy for most homeowners and works with many of the more common perennial flower seeds as well as most annuals. It is also a good choice if you don't have good lights.

If you have indoor lights, the paper towel and baggy method works for a very wide range of seeds. It is also the best method for seeds that require several cold-warm-cold cycles to induce germination.

Very small seeds are more easily done by seeding directly in pots or with winter sowing.

Seeds that are easy and quick to germinate, like most vegetables, are easily done in pots.

Each method has advantages. Give them all a try so that you better understand them. Over time you will gravitate to the one that is best for you and your seeds.

Light Source

Once seeds germinate, they need light to grow and a good light source is the most limiting factor for homeowners growing indoors.

You can grow seeds near a window, but due to space constraints this usually limits the number of seed types you can grow. It is, however, inexpensive and requires no special equipment.

Seeds grow well under artificial lights since they don't need as much light as adult plants. LED technology has advanced greatly in recent years and if you are planning to buy lights, go with LED. You can buy LED shop lights that are inexpensive and great for starting seeds. Even better are LED grow lights specifically designed for plant growth. They are more expensive, but give off much more light.

If you don't have a good light source you can still start seeds with the winter sowing method.

Seed Starting Resources
- 10 Seed Starting Mistakes: https://youtu.be/3QSsX35Z4-g
- Baggy Method for Starting Seeds: https://youtu.be/dirz0WIMQi0
- Growing from Seed (complete set of videos): https://www.youtube.com/watch?v=dirz0WIMQi0&list=PLq7hmpP9i05Ska3k7gaBCvNCT9gN_tYaj
- LED Grow Light Buyers Guide: https://youtu.be/NHJJ-HhinKA
- Winter Sowing: https://youtu.be/SO_KKbGYTEM

11

Selecting Seeds

Gardeners are growing many plants from seed and even collecting their own seed. It is the only way to obtain certain types of unusual plants. Many gardeners become amateur plant breeders and one of the easiest to start with is the daylily. Nothing is as exciting as watching your first hybrid daylily bloom.

Basic Genetics

Plant sex is really no different than animal sex. The mother produces an egg and the father produces a sperm. When these join together and grow, you get a seed.

All higher-level plants and animals have DNA that is a double set of chromosomes. Each chromosome in your body as well as your plants exists as a twin set. When the sex cells (eggs and sperm) are made they get only one half of each chromosome and when they combine to form a fertilized egg, it again has two sets of each chromosome.

What this means is that the offspring of any set of parents has quite a bit of genetic variation and is different from either parent. Even when a single parent is involved (self-pollination) there is still variation.

So why do plants of a single species all look the same in the wild? The reason is that in any given population, the plants have been line breeding themselves for a long time. It is like a small community that has been breeding only within the community with no outsiders involved. Over time this kind of breeding, called line breeding or inbreeding, produces individuals that all look the same.

Plant breeders like to get involved and they will crossbreed plants from different populations or even different species. That introduces a lot of variation in the offspring. Plant breeders can then select the special ones that have the best characteristics and make them available through the industry.

Hybrids vs. Heirlooms

Heirlooms are similar to the wild populations described above. They have been line bred for a long time and over that time all individuals in the population have become similar. If you get some heirloom sugar snap peas they all look the same and produce the same peas.

Hybrids on the other hand are the offspring of crossing two different populations. My favorite tomato is the Sweet 100 and it is a hybrid. The seed is made by crossing two different types of line bred tomatoes. Because the parents are genetically very stable, the offspring are very similar to one another. If you now collect seed from these plants you will get a lot of variation in the next generation.

If you are interested in collecting your own vegetable seed, it is best to grow heirlooms. Their seed will be very stable and produce seedlings like the parents. It is not recommended to collect hybrid seed; you don't know what kind of tomatoes you will get, but most will be inferior to the parents.

GMO Seeds

There has been a lot of talk about GMO plants and seeds and unfortunately the general public has been given a lot of misinformation about them.

I know it sounds very scary to think about taking DNA from one organism and placing it into another, but this has been going on in nature for millions of years. The chloroplasts in plants probably originated as photosynthesising bacteria. Our own DNA is full of virus and bacteria DNA. The sweet potato, for example, contains bacterial DNA. Nature has been mixing up DNA genes from different species for a very long time.

The movement of genes is not new. What is new is that we now have the technology to do it very precisely and in a lab. When done in a lab, we know exactly which genes are moved into a plant.

Now consider what happens in nature or in your backyard. Pollen and egg cells contain thousands of genes. They are all put into a vessel, the flower, and stirred around to produce a seed. There is no control over which genes are combined. There is no evaluation of the potential damages caused by the new seed. We just assume that if nature does it, it must be good, and that is not correct.

Remember that many of the genes in a plant are there to produce toxins. It is not only possible, but has actually happened that naturally bred vegetable seeds have produced plants that have made us sick because of higher toxin levels. After all, breeders select the best plants, and ones that are not bothered by pests make a great selection. These also produce the highest level of toxins.

GMO food is a concern to the general public, but almost all scientists agree that it is a safe way to produce new improved seeds and that GMO food is safe to eat.

Buying GMO Seed

Most garden seed catalogs now make a big deal about the fact that they don't sell GMO seeds. Guess what? GMO seeds are not available to any gardeners unless you buy large amounts and sign a contract. This business of promoting themselves as "non-GMO" is just marketing hype to get you to buy their seeds.

Days to Maturity

The terms "days to maturity" and "days to harvest" are used interchangeably. Days to maturity is the time needed for the plant to reach maturity. That seems simple enough, but when do you start measuring this time period and how is maturity defined?

Maturity is when you can either see flowers, in the case of ornamental plants, or pick vegetables, in the case of vegetable seed. In the case of tomatoes, for example, maturity happens when you can pick the first ripe tomato.

When do you start measuring this time period? The answer depends on how you start the seed. If the seed is usually planted directly in the garden, like carrot seed, then the time to maturity is the time between planting and harvesting the first carrot. If the seed is usually started indoors and then planted in the garden, like tomatoes, the time of maturity is the time from setting out seedlings to the harvest of the first tomato.

Days to maturity is a useful number for gardeners, especially when growing vegetables, but it does have limitations. The point of maturity is a bit ambiguous. For example, I can pick carrots when they are quite small, or wait until late fall. In my zone 5 garden that is a three-month spread. For lettuce some people wait until a head is formed, others pick a few leaves every week.

Location is important. Seed packages have only one days to maturity value, but the same package is sent to many different hardiness zones. Each of these has a different climate, which affects how quickly plants grow. Warmer weather ripens tomatoes faster and temperatures that are too high slow it down. Days to maturity numbers are just approximate for a wide range of growing conditions.

The way to use the days to maturity value is to compare seed packs. They may not be accurate numbers, but they are accurate relative to each other. Radishes have a value of 30 days, and tomatoes have a value of 65 days. I don't expect to harvest the crop in exactly 30 or 65 days, but I do know that the radishes will be ready to pick about four weeks before the tomatoes.

Days to Maturity for Tomatoes

Tomato Variety	Days to Maturity
Beefsteak	96
Early Girl	75
Sweet 100	60

Days to maturity is even more useful for selecting a cultivar. Consider the tomatoes listed in the above table. Sweet 100 tomatoes will be ready to harvest a month before beefsteaks. If you live in an area with a short growing season it is a good idea to pick varieties that have the smallest days to maturity value. You will be able to harvest sooner, and longer, before frost kills the plants.

Buying Unusual Seeds

You are now excited about growing unusual plants from seed but where do you get them? There are many commercial seed companies and this is a good starting point. If they don't have what you are looking for, consider joining specialty societies. Many of these have seed exchanges where members collect seeds and submit them to the organization which then makes them available to members.

If you are new to growing things from seed try some online gardening groups in places like Facebook. Some of these will run seed exchanges for more common seeds. I am still growing some dianthus I got this way.

This link will take you to a list of some societies that offer seed exchanges: https://www.gardenfundamentals.com/seed-exchange-get-free-seed

12
Vegetative Reproduction

Woodlands in northeast North America get covered by an interesting plant called *Erythronium americanum,* the trout lily. You can easily find large colonies of thousands of small mottled leaves and every once in a while one will flower. Only about 0.5% flower in any given year and these plants always have two leaves. A plant that flowers this year may not flower again for several years.

Why are there so many plants when they don't flower very often? The flowering plants do produce some seed but it takes about eight years for these seeds to produce new flowering plants. This is a very slow way to reproduce.

The trout lily solves this problem by also reproducing vegetatively. The mother plant consists of a corm and one leaf. Once the leaf is aboveground, it makes up to three stolons which work their way under the leaf litter until they make contact with the soil and subsequently grow a new plant. Each stolon can produce up to four new plants. Many more plants are produced this way than by seed.

I wonder if the trout lily is on its way to completely replacing sexual reproduction with vegetative reproduction?

There are good reasons for reproducing vegetatively rather than using seed. Producing flowers and seeds are very complex processes that require a lot of energy. It is much more economical for a plant to reproduce vegetatively. The production of stolons is similar to growing stems, making the process easy.

There is however a downside. Genetically, every plant created vegetatively is the same as the original mother plant. The population lacks diversity, and that can be a problem long term. The advantage of sexual reproduction is that offspring are all genetically different, increasing diversity of genes. Such a population is much more likely to adapt to environmental changes.

Natural Vegetative Reproduction

Plants use various methods for vegetative reproduction and it is not uncommon for a species to use more than one method.

What is vegetative reproduction? The previous chapter focused on seeds, which are a form of sexual reproduction—namely a sperm and egg combine, usually from two different parents. The genetics of the seedling is different from either parent.

The term vegetative reproduction is used to describe a process where a piece of a single plant is used to create a whole new plant. The genetics of this new plant is identical to the parent.

From a plant breeder's perspective both forms of reproduction offer advantages. Sexual reproduction is the only way to form new plants, but once they are formed, breeders want them to remain stable so they can sell the identical plant to thousands of gardeners. That is where vegetative reproduction becomes valuable.

Stolons and Rhizomes

Many plants produce stolons and rhizomes. The strawberry makes stolons that root when they touch soil. Quack grass and goutweed produce rhizomes that can root at each node. This is a very effective form of reproduction but most plants that use this method are too aggressive for the garden.

Fragmentation

Fragmentation is a process where pieces of the parent plant break off and grow on their own. Primitive plants such as mosses and liverworts use this method.

A variation of this also occurs in some rhizomes. Initially the plant produces rhizomes to expand the size of the parent plant. Over time these rhizomes break, resulting in separate plants.

Formation of Bulblets, Bulbils and Cormels

A bulblet is a baby bulb that forms underground from a mature bulb. If you dig up tulips or daffodils that have been in the ground for several years, the larger mature bulbs will have formed smaller bulbs at their base.

The term bulbil is used to describe a baby bulb that forms either in a leaf node or a flower. Some lilies will form a bulbil in each leaf node. Garlic forms what looks like a flower bud, but instead of flowers they form baby garlic bulbs, called bulbils.

Corms also form baby corms and these are called cormels or cormlets. If you have ever dug up a gladiolus corm to store it for winter, you will have seen numerous cormels at the base of the corm.

Although each of the above is botanically different, from a gardener's perspective they are all the same. They are baby bulbs that will grow into new plants when put in soil. The genetics of the new plant will be identical to the parent plant.

Bending Branches and Vines

Some plants, like the black raspberry, grow so tall that their arching stems eventually touch the ground where they root and form new plants.

Many vines will root along their stem if it touches the ground. Initially the whole structure remains as one plant, but if the vine gets severed for any reason, separate new plants are created.

Rooted Stems and Leaves

When the stems or leaves of some plants touch the ground, they will form roots that eventually grow into a complete plant. This is very common with some sedum, like acres. Each small piece of plant material will root. This is one reason plants like this are very hard to remove from the garden. Sometimes it is a good idea to remove the plant as well as the soil under it in an effort to get every last bit.

Houseplants like African violets and jade plants root easily if a leaf touches the soil.

In some ferns and the houseplant kalanchoe, tiny plantlets form along leaf margins. These then drop off as complete plants ready to grow.

Apomixis and Why Dandelions Produce So Much Seed

Apomixis is not vegetative reproduction, but it is not sexual either. It is a way in which plants are able to make seed without pollination. An embryo normally contains a half set of genes with the pollen providing the other half during pollination. In the case of apomixis the developed embryo already contains a full complement of genes and does not need the DNA from pollen. This method is used by a number of plants including Kentucky bluegrass, growing in lawns across North America, and many varieties of blackberries.

This method is also used by dandelions. Even when insects are not around, you still get dandelion seeds and the process is quick and simple, which explains why you even get seeds after the flower head has been picked. All of the seed in a seed head is a genetic duplicate of the parent.

Artificial Vegetative Reproduction

Now that you understand what plants can do on their own we'll have a look at methods gardeners can use to produce more plants.

A gardener can use all of the natural methods of reproduction, but can also modify these methods to speed up the process or to control the number of plants produced.

Plant Division for Herbaceous Plants

This is perhaps the easiest method. Many perennial plants can simply be sliced into smaller pieces with each piece growing into a new plant.

This works well for any plant that forms a crown that is made up of stolons, rhizomes or tubers. A bearded iris grows using rhizomes and these can be broken into smaller pieces. Provided each piece has a developing bud, the rhizome will grow into a new plant.

Dahlia tubers can be split down the old stem ensuring that each piece has some dormant eyes that are located at the base of the stem.

If you look at a clump of daylilies or hosta you might see one large plant but it is really numerous individual plants joined together by a common root system. You can break such clumps into individual plants provided that each division has an active bud or leaf and its own roots.

In practice it is best to make larger clumps. These will grow more easily and look much better in the garden. I generally divide a large clump into four pieces of approximately equal size and the best time to do this is during the cooler part of the year. In temperate climates spring and fall work best.

Plant division does not normally work well for woody plants because their crown does not respond well to being split. This is also true for most subshrubs like lavender and Russian sage.

Layering

Layering is a method where a stem is bent down to the ground and covered with soil. It is best to have at least one node under the soil since this is the most likely place that roots will form. The tip of the stem remains above the ground so its leaves can continue to grow.

I usually place a good-sized rock on the stem to hold it in place, and to help keep moisture in the soil.

The process works well with woody plants, but it can also be done with any herbaceous plant that makes stems which can be easily bent down. It is also a great method for producing new plants from vines like clematis.

Layering is easy, and takes no maintenance. It is a good method if you only need to produce one or two plants—perfect for most home gardeners. It is limited by the fact that you need to find a stem that can be bent to the ground. If you can't find one, you might have to start training a more vertical one so that it is more horizontal, and eventually train it to grow down.

After 6 to 12 months, depending on when the stem is layered and the type of plant, you can move the rock and gently uncover the

Layering a branch.

stem to confirm it has rooted. If you don't see roots, bury it again and wait. If the stem has rooted, cut it between the mother plant and roots. Pot it up or plant it directly in the garden.

A modified layering method can also be used for difficult-to-root plants. Before burying the stem, take a knife and scrape the bottom of the stem where it will be buried. Some people even cut a sliver of stem right off. This exposes the cambium layer, the part of the stem that roots more easily.

You can also put on some rooting hormone, but to be honest, I never use it when layering.

Leaf Cuttings

Many gardeners start learning about vegetative reproduction by taking stem or leaf cuttings from houseplants. Many of these, including coleus and money plant, root easily and the most common method is to simply place the plant material so that the lower section is in water. Once they form roots they can be potted in soil and treated like the mother plant.

The use of water is very common and it does work in a lot of cases, but it is not the best option. For one thing, some plants just rot in water. Secondly, plants that form roots in water grow special water-adapted roots. When you then move the plant to soil, it needs to grow a new set of roots. Why not just root them in soil to start with? You can use the same method described below for stem cuttings.

The Magic of Stem Cuttings

I have reproduced many plants by making cuttings, and every time I do, I am still amazed that it works.

On the surface the method is quite simple. Take some pieces of plant material, usually stems, but they can also be roots. Place these cuttings in a humid environment and wait. After a while, the cuttings form roots, followed by shoot growth. Pot up the rooted cutting and you have a new plant.

Many of the plants you buy are created this way and commercial operations usually have greenhouses and misting systems. They get a high rate of rooting. This sounds very complicated, and it is, but it does not have to be. The method I will describe is one I developed for the average gardener who has no equipment. It is a simple method that will result in roots on about half of the cuttings, depending on the species you try. I use it mostly for rooting stem cuttings.

There are four basic types of stem cuttings; herbaceous, softwood, semi-hardwood and hardwood.

Stem cuttings can be taken from many types of herbaceous plants and they root best if taken from young stems.

A softwood cutting is one taken early in the growing period. The stem is still green and very flexible.

A semi-hardwood cutting is taken in mid-season. The growth of woody plants has now mostly stopped and the stem starts its hardening off process. The color changes from green to a light brown, and the stem is stiffer.

Hardwood cuttings are taken in fall or even winter and are fully hardened off. They are dark in color and have very limited flexibility.

The three types of wood are described as if there are distinct phases of development, but that is not true since the hardening off process is a slow process that happens over months. A cutting taken in late spring is halfway between soft and semi-hard. The other thing you notice is that in late spring or early summer, the tip of the branch might still be softwood, but the lower portion is already semi-hardwood. Don't get hung up on finding the exact perfect time to take the cuttings.

Which type of cutting works best? That depends on the plant. Many herbaceous perennials propagate easily from very early softwood cuttings. Most shrubs root easily from both softwood and semi-hardwood cuttings. French lilacs can be difficult to root, but if you take the cutting just as the flowers are finished, most are quite easy to root. Korean lilacs on the other hand can be rooted easily almost any time.

Evergreens can be easy or difficult to root. Pines are almost impossible, no matter which type of cutting you use. Boxwood and yews are child's play and you can take cuttings in spring or summer. The Sawara cypress (*Chamaecyparis pisifera*) is not easy to root unless you take cuttings in late winter. Most deciduous trees are hard to root.

I know this all sounds very complicated. You can look up each species online and find out the type of cutting that works, or you can follow my simple rules that will work most of the time:

- Trees are difficult to root.
- Shrubs are easy to root with softwood cuttings, and almost as easy with semi-hardwood cuttings.
- Herbaceous plants root easily as very young softwood cuttings.
- Most evergreens are difficult to root.

If you are new to this, try some of the easy-to-root plants to learn the technique and do several different types of plants.

When I am trying to root a new species I employ a few tricks to increase my success rate:
- Take cuttings every 30 days from mid-spring until late summer. Some will do better than others.
- If a shrub is making long growths, use a full stem in mid-summer. The tip is softwood and the lower portion is semi-hardwood. Try rooting both ends and the piece in the middle.
- Propagate from Cuttings Video: https://youtu.be/DJuUQq0GiFU

My Rooting Process
The first step is to prepare a small rooting greenhouse. Find a clear plastic dome. A large pop bottle works or you can use an old juice bottle. You want something that is at least six inches tall. Cut the bottom off.

Take a standard flower pot that is large enough so the dome fits inside it. Fill the pot with rooting soil. You can use a peat-based potting soil, straight peat moss, perlite, vermiculite or even sand. I like to mix half potting soil and half perlite to create a very airy medium,

Semi-hardwood stem cutting from a lilac branch.

but in a pinch I have even used garden soil. Roots will just grow more slowly in heavy soil.

Select your plant and remove several branches while it is cool: early morning or evening. Process the cuttings right away. A cutting should have two things: a node at the bottom and a leaf at the top.

I make a diagonal cut right below a node. The diagonal cut reminds me that this is the bottom of the cutting; like a spade, it goes into the soil. Cuttings will not root upside down.

I then make a horizontal cut right above a leaf. If there are two leaves at this node, remove one and if the remaining leaf is very large you can cut half of it off.

The very tip of a branch, the point where leaves are still growing, is usually not used for woody cuttings but it can be used for very early herbaceous cuttings.

The length of the cutting should be about three inches but the actual size depends on the source material. A cutting may just be one internode long, or it can have several internodes.

I usually make six to eight cuttings per propagation pot. That usually gives me three or four rooted plants, plenty for the average gardener.

Take each cutting and dip it into commercial rooting hormone. Don't use too much. Higher levels of hormones prevent the formation of roots. Dip the cutting in the powder and tap it on the side of the container to knock most of the hormone off. I prefer using the powder because it has a much longer shelf life than the liquid form. Ignore the expiration date on bottles—dry rooting hormone lasts at least five years and probably ten.

Is the rooting hormone necessary? Most plants will root without it, but it may speed up the process. If you have it, use it. If you don't have any, try without it.

Now stick the cuttings in the rooting soil so the leaf is just showing above the soil level. Water, and cover with your plastic dome. Set the pot in a shady spot. Keep the pot watered, but you will find that it does not dry out very quickly. Don't keep it too wet.

Open it from time to time and remove any rotting leaves. Even if a cutting loses the leaf, there is still a good chance it will root. Remove any cutting that is rotting or covered in mold.

Many gardeners new to this process get all excited when they see new growth at the dormant bud. If this happens soon after sticking the cuttings, it is actually not a good sign. The cutting has prioritized stem and leaf growth over root growth and such cuttings almost always fail. What you really want to see is nothing, for several weeks. All the important growth is happening underground—the focus has to be on the roots.

> ### Plant Myth: Homemade Rooting Hormones Work Well
>
> Commercial rooting hormone contains plant hormones, it works well, and since a bottle of powder lasts ten years it is a good buy. I don't understand why gardeners want to use much less effective home remedies.
>
> Here are some common DIY rooting hormones and their effectiveness:
>
> - Aspirin—this is not a rooting hormone, but low amounts may help roots grow once formed. It can become toxic quickly.
> - Cinnamon—prevents fungus from growing, but has no effect on rooting.
> - Peroxide—sterilizes cutting and may help roots grow more quickly once formed, but it does not seem to help initiate roots.
> - Willow water—willow stems do contain rooting hormone and an extraction of these will provide a very low level of hormone for rooting. It may help, but the science does not support its use. Is it worth making willow water every time you want to root something?
>
> Some other DIY solutions include aloe vera juice, vitamin C and apple cider vinegar. None of these have any scientific evidence to support their use.

After a few months you can give the cutting a slight tug. If there is resistance you might have roots. Once you think the cuttings have rooted, empty the pot and check the cuttings. If they have roots, pot them up individually and care for them just like any other plant. The plastic dome is no longer needed.

Grafting

Grafting is a process where two or even three different types of plants are connected together to form a new plant. Most fruit trees are grafted and in the past most roses were also grafted. More roses are now available on their own roots.

Grafting can be a quick way to create many copies of a plant, especially for plants that don't root easily by other means. For example, apple trees are difficult to root by stem cuttings but gardeners want to grow a particular cultivar. They can't be reproduced by seed because the seedlings will all be different. Grafting is a way to make many plants that are genetically identical.

Grafting is also used when a plant's root system is weak, or not well adapted to a particular environment. Growers will take a very hardy, disease-resistant root system and graft a more favorable cultivar on it. You get the best of both worlds: a good root system and great flowers and fruits. For example, many named cultivars of witch hazel are grafted onto wild native root stock.

The technique is also used to grow plants that never have any roots of their own. Most dwarf conifers do not exist in nature as trees. These miniature growths develop on normal trees as witches' brooms. Cuttings of these will not root on their own so you can't form new trees. The solution is to graft a piece onto a wild seedling. Once the graft forms, you have a dwarf tree with a root system.

Many tree peonies are also grafted but they are unusual grafts. The tree peony cutting is grafted onto a root from a herbaceous peony. The new plant is a combination of the two types of peonies, but the aboveground portions will look and grow like the tree peony. Over time the tree peony will form its own roots and abandon the so-called nurse root.

The Grafting Process

Grafting is a simple process but it does take practice to get it right. In simple terms a scion, which is a piece of the top growth, is connected to the trunk of a root stock. There are several ways to make this connection but it is important that the vascular systems of both pieces line up. If done properly, they fuse together and form a fully functional union. Any new growth from the root stock is removed so only the scion grows.

If you are new to grafting and thinking of giving it a try, start with bud grafting, also called budding. It is less technical and can be done in mid-summer.

Grafted Trees

Most gardeners don't graft their own plants but they do buy grafted plants so it is important to know how to maintain them.

The first step is to identify the graft union. Most purchased trees are still fairly young and the graft is quite visible as a swollen

Graft union on a weeping beech.

point, usually near the ground. I have a special prunus that is double grafted. It has one graft a few inches above the soil line and a second one at about five feet high. It consists of three different plants but such trees are not common.

The grafted tree gets planted and maintained just like any other tree except for suckers. Grafted trees have a tendency to sucker from below the graft. Since this growth is from the root it will not have the desirable characteristics of the top growth, so it must be removed.

The common approach to removing suckers is to prune at ground level, but all that does is produce more suckers the following year. Remember, these suckers have dormant buds on them and you need to remove them too. The best thing to do is dig down until you find the origin of the sucker. Now cut as close to the trunk or root as you can. I have even scraped the growth away to try and kill any dormant buds right at the joint.

As the upper part of the tree gets bigger, many grafted trees will stop suckering, but if they don't you have to keep after them each year.

The same problem exists for tree peonies. If they sprout from the herbaceous peony root you will have two kinds of peonies growing. The growth from the root is almost always less desirable since it is selected for its root quality, not its flower. The leaves of the two peonies are quite different and you should have no trouble telling them apart.

13
Plant Names

Knowledge of plant names is the foundation of gardening and gardeners use it to select plants and determine how to grow and maintain them. Many gardeners obtain or share knowledge by communicating with other plant enthusiasts from around the world and a good grasp on plant names will save significant time, effort and money.

A rose by any other name might not smell as sweet…because it's an entirely different plant!

Why Use Botanical Names?

You may have noticed that gardeners who have several years of experience or formal training tend to refer to plants by their botanical names. Newer gardeners, or gardeners who take a less technical approach, often refer to plants by their common names.

Common names sound interesting and are easy to remember. However, common names are completely unregulated, unlike botanical names, so there's a good chance that another person will assume you are referring to a completely different plant. Using a plant's scientific name increases the odds that you're buying the correct plant and employing the correct methods for planting, growing, maintenance and propagation.

A good example of a confusing common name is the outhouse plant or outhouse flower. Before the days of indoor plumbing, tall flowers were grown around outhouses to disguise their look and smell. *Rudbeckia laciniata* 'Hortensia' a tall perennial with double yellow daisy flowers, is the plant that most have in mind when referring to the outhouse plant. Yet, in online gardening groups, outhouse plant is used to refer to other tall yellow plants that resemble daisies. If you drop the 'Hortensia' from *Rudbeckia laciniata* you get the native cutleaf coneflower that boasts large single yellow daisy flowers and a prominent grey-green pistil. Hollyhock (*Alcea rosea*), a tall but short-lived biennial, is also called outhouse plant because it was also commonly planted around outhouses. It would be a major shock to order "outhouse flower" expecting clumps of fluffy yellow *Rudbeckia laciniata* 'Hortensia' only to receive straight and stately *Alcea rosea*.

Many of the online requests for information provide the wrong advice simply because people are using common names. It leads to a lot of confusion.

Naming Conventions

Plants can either have a common name and a botanical name, multiple common names and a botanical name, or a botanical name but no common name (or at least not one that is well-known). Sometimes plants share common names with other plants, but a scientific or binomial name can only belong to one species.

The common name is simply the name that caught on in a certain area over time. In contrast, the botanical (or binomial) name corresponds to a precise and universal system—the International Code of Botanical Nomenclature, developed by 18th century botanist Carl Linnaeus. The binomial name consists of two italicized words, the first corresponding to the plant's genus and the second to the species. A genus is a grouping of species with similar characteristics, whereas a species is made up of individuals that share specific traits, and can reproduce and make viable (i.e., fertile) offspring.

To give an example, *Cercis canadensis* is the botanical name for the eastern redbud, also known as the American redbud.

Only the genus name is capitalized when using the binomial system. Words in the common name are generally not capitalized unless they are a proper noun, but sometimes capitals are used as a stylistic choice.

Even with the binomial naming system there are limitations. Plants are often recategorized and renamed as genetic research progresses. This was the case for New England aster, *Symphyotrichum novae-angliae*, which was changed from *Asteraceae novae-angliae* after it was discovered that it's not related to the aster family despite the similar flower shape. Adding to the confusion is a plant's tendency to produce hybrids or develop new traits through mutations. This brings us further nuances in plant names.

Naming Hybrids

Primary hybrids are a cross between two different species. They can occur in nature and be the result of crossbreeding.

There are two ways to name hybrid plants using the binomial naming system. Firstly, if the hybrid is well-known enough to have been given a specific name, then the name will be the genus, followed by "×" and the hybrid name. For example, *Acer × freemanii* is the binomial name of the Freeman maple, a naturally occurring cross between *Acer saccharinum* and *Acer rubrum*. Alternatively, the hybrid could also be called *Acer rubrum × saccharinum*.

Plants are more promiscuous than other forms of life, so there are also cases of hybrids at the genus level, called intergeneric hybrids. It is common to name an intergeneric hybrid after a combination of the two genera names, such as with *Sorbaronia* shrubs, which are a hybrid between *Sorbus* and *Aronia*, and formally called × *Sorbaronia*.

What about Varieties and Cultivars?

It's common for gardeners to use "variety" and "cultivar" interchangeably or incorrectly when speaking about plants.

The main difference between a variety and a cultivar is that a variety occurs naturally in the wild and a cultivar is manmade. A variety will produce seeded offspring with the same genetic makeup as the parents and these may be propagated in nurseries. In the case of cultivars, many do not breed true and need to be propagated vegetatively. In the case of some crops, like oats and wheat, we have been line breeding these cultivars for so long that they can be propagated by seed and still have the same genetics as the parents. Hybrid seeds, like the Sweet 100 tomato, are also cultivars.

Cultivars and varieties also have different naming conventions. To give an example, *Cercis canadensis* var. *alba* is the name given to a variety of eastern redbud with white flowers. White flowering redbuds are naturally occurring—the plant produces white flowers rather than red due to a mutation. Cultivars, on the other hand, are named with the cultivar name after the botanical name in single quotations, for example, *Cercis canadensis* 'Appalachian Red'. The common name would then be Appalachian Red eastern redbud. It's also acceptable to use the name *Cercis canadensis* cv. Appalachian Red, but this is rarely used.

Lastly, it is also possible to produce cultivars from varieties. In that case, the botanical name would be the variety name followed by the cultivar name in single quotations; *Cercis canadensis* var. *alba* 'Appalachian White'.

The Proper Way to Name Your Plants

Plants are highly productive and will make new plants for years to come, whether from seeds or vegetative structures. For wild or cultivated varieties, the naming process is quite simple: keep the same binomial name since the genetics have not changed. *Cercis canadensis* var. *texensis* grown from self-pollinated seed is still *Cercis canadensis* var. *texensis*.

Naming becomes more complex with cultivars. If you propagate a cultivar asexually then the name can remain the same since it is the same plant, with the same genetics.

Cultivars grown from seed, including seedlings you find growing near the parent plant, can be given the binomial name, but the cultivar name should not be used because you can't guarantee the new plant has the same genetics. Even if the mother plant was self-pollinated, the seedling now has different genetics and should not be given the same cultivar name.

If I want to keep track of a particular seedling, I will give it my own cultivar name. This allows me to keep track of it, and when someone else gets the plant they won't confuse it with accepted named cultivars. So you might find an orchid out there called *Phalaenopsis* 'Poopy' which I developed years ago, or a more recently developed lilac called *Syringa* 'Judy', which I named after my wife. Neither plant is registered.

If you have an unnamed plant, don't give it a cultivar name you find online or in a magazine. This just leads to more future mistakes when you share the plant with friends. The names in the nursery industry are already confused enough. I have three *Phlox paniculata* 'David' in the garden, all different, because people assign the name David to any white flowering summer phlox.

Index

Page numbers in *italics* indicate tables and illustrations.

A

abiotic stresses, 70
abscisic acid, 166
abscission zone (AZ), 69
acclimation, 149, *150*, 151–152
active transport, 27–28
adaptability of plants, 149, *150*, 151–152
ADP (adenosine diphosphate), 10–11
adventitious roots, 21
African violets, 148, 184
agave (century plant), 81, 104
All-American Trial Gardens, 140
allelochemicals, 38–39
alpine plants, 115–116
amino acids, 7
ammonium, 109
anemones, 49
angiosperms, 14, 77, 160
annuals, life cycle, 101–102
anthers, 76–77
anthocyanins, 62–63
anthurium, 88
apical buds, 44, 45, 127–130
apical dominance, 127–128
apomixis, 184
apple trees, 93, 95, 192
ash trees, 130
asparagus, 43
aspen, 150, 151–152
ATP (adenosine triphosphate), 10–12
auxins, 38, 127–128
axillary buds, 44, 45
Azospirillum bacteria, 38

B

bacteria, 32–35, 36–37
baggy seed starting method, 170–171
bamboo, 44
bananas, 95
bark, 120–121, 124–125, 145
bats, 86
beans, 34, 38, 53, 93, 106
bearded iris, 52, 184
bees, 85, 86
beets, 102
begonias, 102
bell peppers, 158
Berkheya purpurea, 140
biennials, life cycle, 102–103
black raspberries, 183
blackberries, 184
bloom duration, 79
blooms. *See* flowers
blossom end rot, 112
blue festuca, 107
blue poppy (*Meconopsis*), 103–104, 139
boron, 109
botanical names, 195–199
bougainvillea, 87
boxwood, 188
bracts, 87–88
branch collar, 131, *132*, 133
branches, 117, *121*
breaker stage, 95
breeding, 175–176
buds
    apical dominance, 127–130
    on conifers, 133–134

meristem tissue, 12–13
plant damage, 47, 114–115
types of, 44–45, 131
in winter, 82, 125
bulbils, 183
bulblets, 183
bulbs, 21, 49–51, 51, 81–82
butterflies, 86

C

cacti, 48, 59, 146, 147–148
calcium, 7, 26, 109, 112
CAM (Crassulacean Acid Metabolism) plants, 59
cambium layer, 120
Canada, hardiness zones, 138
carbohydrates, 7, 10
carbon dioxide, 8–9, 10, 57–58, 154
carrots, 20, 102, 178
celery, 6
cells, structure, 5
 See also meristematic cells
cellulose, 5
century plant (agave), 81, 104
chelates, 38, 40
chili peppers (*Capsicum annuum*), 158
chloroplasts, 9, 57
chlorosis, 110
Christmas cacti, 80
chrysanthemums, 46, 79
circulating roots, 22, 23
classification of plants, 14
clay, 26
clematis, 45, 53, 88, 119, 185
climate change, 10, 150, 152–156
climbing plants, 21
clochicum, 148
coconut trees, 44
cold temperatures, 140–145
Colorado blue spruce (*Picea pungens*), 56
common poppy (*Papaver rhoeas*), 98
concorde barberry (*Berberis thunbergii* 'Concorde'), 62
coniferous trees, 59–60, 130–131, 133–135

container plants, protection, 142–143, 145–146
containers, seed starting, 169–170, 171–173
copper, 109
coppicing, 129
cork cells, 120–121
cormels, 183
corms, 50, 51–52
corn, 158
cortex, 17, 43–44
cosmos, 102
cotyledons, 160–162
crab apples, 81
creeping phlox (*Phlox stolonifera*), 139
crocus, 51
cucumbers, 53, 94, 106, 158
cultivars, naming conventions, 197–198
cuticle, 60
cytokinins, 38

D

daffodils, 51, 183
dahlias, 21, 47, 49, 185
damage, 114–115, 131–133
dandelions, 20, 115, 184
Darwin, Charles, 98
day length, 79
daylilies, 48, 185
deadly nightshade, 92
deciduous trees, 105, 107
determinate plants, 105–106
dicots, 14, 43, 160
differentiated cells, 12
dioecious plants, 88–90
disease, 64, 146
DNA, 175–177
dormancy, 106–107
dormant buds, 12–13, 44–45, 47, 114–115, 131, 133–134
double dormancy, 164
Douglas fir, 160
driplines, 20
drought, 147–149
drought-tolerant plants, 59–60, 147, 155
dwarf conifers, 192

E

eggplants, 158
energy, 7–12
environmental factors, 137–156
epicormic buds, 130–131, 134
epidermis, 17, 43, 56–57, 101
espalier, 128, 129–130
ethylene, 95
evening primrose, 86
evergreen trees, 25, 26, 63, 105–106, 142–143, 188
exudates, 36–37, 39–40

F

fabric plant containers, 145–146
fall changes, 63, 70, 109–110, 113
ferns, 14, 184
fertilizer, 3–4, 27, 82–83, 155
fibrous roots, 19–20, 125–126
filaments, 76–77
flowering dogwood, 88
flowers
    bloom prevention, 84–85
    blooming criteria, 78–84, 116–117
    bracts, 87–88
    buds, 44
    monoecious and dioecious, 88–90
    pollination, 77–78, 85–87
    structure, 75–77, 76
foliar feeding, 111–112
forsythia, 44, 45, 82
foxgloves, 102
fragmentation, 182–183
free-living nitrogen-fixing bacteria, 33
frost damage, 141–142
fruit set, 93–94
fruit trees, 79, 130, 192
fruits, 91–95
fungi, 31–32, 64
fungus gnats, 86–87

G

gardening zones, 137–140, 155
garlic, 144, 183
garlic mustard (*Alliaria petiolate*), 39
genetics, 175–176
German bearded iris, 76
germination inhibitors, 165

gibberellic acid, 166
ginkgo (*Ginkgo biloba*), 89–90, 98
gladiolus, 51, 183
glucose, 9
GMO seeds, 176–177
goldenrod, 44
grafted trees, maintenance, 193–194
grafting, 192–193
grape vine, 53, 119
gravity, 21–22, 117–118
grow lights, 174
growing conditions, 137–156
growth, meristematic cells and, 13
guttation, 108–109
gymnosperms, 14, 160

H

hardiness zones, 137–140, 155
hardwood, 122–123, 187–188
Harlequin maple (*Acer platanoides* 'Drummondii'), 72
heartwood, 120
heat, 145–146
heather, 120
hedges, 130
height, 115
heirlooms, 176
hellebores, 88
Helmont, Jan van, 127
hens and chicks (sempervivums), 104
herbaceous plants
    CO_2 capture, 10
    definition, 5
    dormant buds, 44–45, 47
    stems, 41, 45–46, 140
    vegetative reproduction, 184–185, 187–188
    woody plants vs., 119
high bush cranberry, 93
holly, 89
honeysuckle, 53
hormones, 7, 38, 166
    See also rooting hormones
hostas, 44, 67, 72, 185
houseplants, 80, 117, 148
hummingbirds, 86
hyacinths, 51

hybrids, 176, 197
hydrangea, 53, 82, 139
hydrogen ions, 26

I
imbricate bulbs, 51
impatiens, 102
imperfect flowers, 75
indeterminate plants, 105–106
inner bark, 120, 124
insecticidal soap, 60
insects, 154
internodes, 42
interveinal chlorosis, 30, 110
invasive species, 153
iron, 26, 40, 109, 110
irrigation, 148–149
ivy, 21

J
jack-in-the-pulpit (*Arisaema triphyllum*), 86–87, 90
jade plants, 184
Jiffy pellets, 169–170
juglone, 38–39
junipers, 90

K
kalanchoe, 184
Kentucky bluegrass, 184

L
lateral buds, 44, 45
lavender, 120
lawns, 27, 106–107
layering, 185–186
leaf cuttings, 186–187
leaves
    abscission, 68–71
    bracts, 87–88
    damage and disease, 63–64, 114–115
    effect of light, 60–62
    nutrient deficiency, 110
    nutrient movement, 109–110
    protection from pests, 64–67
    red colored, 62–63
    resource sharing, 113
    rooting, 183
    structure, 55–60, 56, 57
    sugar movement, 112–113
    variegation, 71–73
    water stress, 67–68
LED lights, 80, 174
legumes. *See* beans; peas
lenticels, 121
lettuce, 27
life cycles, 101–107
light
    effect of, 47–48, 60–62, 79, 80, 116–117
    grow lights, 174
    for seed germination, 165–166
    variegation and, 72–73
light bulbs, 80
ligularia, 68
lilacs, 82, 84, 188
lily, 76
lily of the valley (*Convallaria majalis*), 52
liverworts, 14
long night plants, 79

M
magnesium, 26, 109
magnolia, 45, 82
maidenhair fern trees (*Ginkgo biloba*), 89–90, 98
malate, 59
manganese, 26, 109
marigolds, 102
marijuana, 89
mast years, 83–84
meristematic cells
    about, 12–13, 13
    damage and, 114
    lateral growth, 123–124
    in root growth, 19, 22
    spring stem growth, 45
mesophyll cells, 57–58
Mexican fireweed (*Bassia scoparia*), 39
microbes, 31–38, 64
migration, 151–152
milkweed, 64
minerals, 7, 39–40
miscanthus, 107, 153

monocarpic plants, 104–105
monocots, 14, 43, 44, 160
monoecious plants, 88–90
mosses, 14
moths, 86
mulch, for cold protection, 144
mycorrhizal fungi, 31–32, 37

N
naming conventions, 195–199
native plants, 156
natural pesticides, 64–66
nectar, 85
needle drop, 134
negative ions, 26–27
nematodes, 36, 38
night, 9, 79–80
night neutral plants, 79
nitrate, 26–27
nitrogen, 27, 82, 110
nitrogen-fixing bacteria, 32–35
nodes, 42
Norway maple (*Acer platanoides*), 72, 159
no-till methods, 154
nursery potted plants, 15, 23, 84–85
nutrients
    absorption of, 26–28
    attraction of, 39–40
    deficiency, 110–111
    definition, 4
    effect of pH, 30
    leaf application, 111–112
    movement through plant, 109–110
    toxicity, 31

O
oak trees, 63, 81
onion, 51
orchids, 83, 85, 148
organic matter, 3, 26
ornamental grasses, 48, 107
orthodox seeds, 157–158, 167–168
osmosis, 27, 108
outer bark, 120–121, 125
ovaries, 77
oxygen, 8, 11–12, 58

P
palisade mesophyll cells, 57
palm, 70
pansies, 102
paper towel seed starting, 170–171
pawpaw (*Asimina triloba*), 89
peas, 34, 38, 53, 93, 106
penstemon, 103
peonies, 21, 45, 103, 131, 161, 164
peperomia, 73
peppers, 93
perennials
    blooming problems, 81
    dormant buds, 131
    in dry conditions, 147
    lateral bud growth, 46
    life cycle, 103–104
    subshrubs vs., 120
    from temperate regions, 102
perfect flowers, 75, 88
pesticides, natural, 64–66
pests, 146, 154
petals, 75, 76
petunias, 102
pH, 28–31, 30
phloem
    in leaf abscission, 69
    role of, 6–7, 109, 112–113
    in roots, 17
    in whole plants, 101
    in woody plants, 120, 124
phosphate, 24, 26, 109
phosphorus, 31, 39–40, 63, 81
photosynthates, 10
photosynthesis, 7–10, 48–49, 57, 58
phytochromes, 79
pineapple sage, 80
pistils, 75, 77
pith, in stems, 44
plant division, 184–185
plant names, 195–199
plants
    choosing diversity, 155–156
    classification, 14
    hardiness zone assignment, 139, 140
poinsettia, 87
pole beans, 45, 53

pollarding, 129
pollination, 77–78
pollinators, 85–87, 154
positive ions, 26
potassium, 81, 109
potatoes, 21, 38, 45, 49, 106
primary growth, 123
proton pumps, 26
protozoa, 36
pruning, 84, 128–130, 132, 133–134

Q
quackgrass (*Elymus repens*), 53

R
radicle, 160
radishes, 178
rainfall, 79, 146–148
raspberries, 93
recalcitrant seeds, 158, 159, 167–168
redbuds, 105, 139, 198
reproduction. *See* flowers; seeds; vegetative reproduction
resource sharing, 113
respiration, 10, 11–12
resurrection plants, 149
Rhizobium bacteria, 33–35
rhizomes, 52, 182–183
rhizosphere, 35–40
ricin, 65–66
riparian plants, 155
root caps, 18, 19, 22, 35, 37
root cuttings, 47
root hairs, 17–18
root stimulator solutions, 24
rooting hormones, 24, 190–191
roots
    absorption process, 7, 26–27, 28, 30
    of annuals, 102
    conditions affecting growth, 21–26, 35, 149
    damage, 22–23, 114–115
    dripline and, 20
    exudates, 36–37, 39–40
    health of, 15–16
    microbes and, 31–38
    oxygen requirements, 11–12
    resource sharing, 113
    role of, 16–17, 112
    structure, 16, 17–18
    types of, 19–21
    in woody plants, 125–126
rose cones, 144
rose of Jericho (*Selaginella lepidophylla*), 149
roses, 79, 143–144, 192
Royal Horticultural Society, 138
rubber trees, 79
runner beans, 53
runners, 21, 52–53, 182

S
sage, 120
sapwood, 120
Sawara cypress (*Chamaecyparis pisifera*), 188
scarification, 166
science, advantages of, 1–2
secondary growth, 123–124
sedum, 183
seed exchanges, 179
seed starting
    choosing seeds, 176–177
    indoors, 168–171
    lighting, 47, 173–174
    root growth, 21
    winter sowing, 171–173
seeds
    of annuals, 102
    cotyledons, 160–162
    days to maturity, 177–179
    development of, 95–97
    dispersal of, 92
    dormancy, 162–167
    germination, 159–160
    maturity, 157–159
    microbes and, 37
    from non-flowering plants, 98
    soil seed bank, 98–99
    storage, 167–168
    types of, 176–177
semi-hardwood, 122–123, 187–188
sepals, 75, 76, 88
shade, effect on growth, 116–117
shoots, 41

short night plants, 79
short-lived perennials, 103
shrubs. *See* woody plants
signaling compounds, 66
snapdragons, 102
snowdrops (*Galanthus nivalis*), 159
soap sprays, 60
softwood, 122–123, 187–188
soil
    adding mycorrhizal fungi, 32
    after drought, 147
    after heavy rainfall, 146–147
    heating effect of, 141
    mounding, 143
    nutrient availability, 28–31, 30
    over-fertilization, 27
    oxygen requirements in, 11–12
    root growth, 19, 24–25
    tilling, 154
soil seed bank, 98–99
soybeans, 111, 158
spongy mesophyll cells, 57–58
staking trees, 124
stamens, 75, 76–77
stem cuttings, 187–192
stems
    damage, 114
    effect of light, 47–48
    functions, 41
    growth, 45–46, 115
    overwintering approaches, 46
    rooting, 183
    specialized, 52–53
    structure, 42–43, 42, 43
    underground, 49–52
    in woody plants, 120–123
stigma, 77, 78
stolons, 21, 52–53, 182
stomata, 43, 58, 111
strawberries, 21, 53, 93
streptocarpus, 148
styles, 77
suberin, 120–121
subshrubs, 120, 185
succulents, 21, 59, 104, 146, 147–148
suckers, 194
sugar maples, 105, 125
sugars, 7, 8–10, 49, 112–113, 125

sulfate, 26, 109
sun, following, 115–116
    See also light
sunflowers, 102, 115–116
super-cooling, 142
sweet potatoes, 21, 176
sycamore (*Acer pseudoplatanus*), 159
symbiotic nitrogen-fixing bacteria, 33–35
synthetic pesticides, 65

T
taproots, 20, 125–126
temperature
    coping with extremes, 140–146
    hardiness zones, 137–139, 138
    for seed germination, 166–167
    See also climate change
tendrils, 53
tepals, 76
terminal buds, 44, 45, 127–130
thyme, 120
tilling, 154
tomatoes
    calcium foliar spray, 112
    days to maturity, 178–179
    determinate vs. indeterminate, 106
    fruit maturation, 94, 95
    germination inhibitors, 165
    insect defense, 57
    seed maturation, 158
    suckering, 46, 97, 97
    upside-down plants, 118
    viviparous germination, 96–97
topiary, 130
transcuticular pores, 111
transplant solutions, 24
tree peonies, 192, 194
trees
    CO_2 storage, 10, 154
    composition of, 127
    driplines, 20
    girdled trunks, 133
    grafted, 193–194
    height of, 115
    planting, 124
    root growth, 21–22

shapes of, 117
wrapping, 142–143, 145
See also woody plants
trembling aspen (*Populus tremuloides*), 150, 151–152
trial gardens, 140
trichomes, 44, 56–57
trilliums, 103, 147, 164
tropical plants, 79, 107
trout lily (*Erythronium americanum*), 181
trunks, 133, 145
tuberous begonias, 21
tuberous roots, 21
tubers, 49
tulips, 50, 51, 183
tunicate bulbs, 51
twiners, 53

U
UK, hardiness zones, 138
undifferentiated cells, 12
United States Department of Agriculture (USDA), 137–138
USA, hardiness zones, 137–139

V
variegated leaves, 71–73
varieties, naming conventions, 197–198
vascular bundle, 6, 17, 43, 101
vegetable gardens, 144
vegetative reproduction, 181–194
vermiculite seed starting method, 171
vernalization, 81–82
vines, 45, 53
viviparous germination, 96–97

W
walnut trees, 38–39, 166
water
  excess, 146–148
  irrigation, 148–149
  low levels, 59–60, 67–68, 147–148
  movement through plant, 6–7, 58, 107–109
  in photosynthesis, 8–9
  roots and, 26, 27, 35
  for seed germination, 164–165
water sprouts, 130
watermelon, 158
weather, abnormal, 152–153
  See also climate change; temperature
weed seed bank, 99
weeping trees, 128
white trillium, 167
whitewashing trunks, 145
wind, 134–135
wind pollination, 85
winter, 25–26, 46, 82, 134–135
winter sowing, 171–173
winter sunscald, 145
winterberry (*Ilex verticillata*), 89
wisteria, 53
witch hazel, 192
woody plants
  buds, 127–131
  damage, 131–133
  formation of wood, 123–125
  hardiness, 139, 140–141
  meristematic cells in, 13
  over-fertilization, 27
  pruning, 84
  stems, 41, 120–123
  sugar storage, 125
  types of, 119–120
  vegetative reproduction, 185, 187–188
  wrapping, 142–143, 145

X
xylem
  function, 6–7, 101, 108
  in leaf abscission, 69
  in roots, 17
  in woody plants, 120, 124

Y
yews, 134, 188
Yucca filamentosa, 104

About the Author

ROBERT PAVLIS, a Master Gardener with over 45 years of gardening experience, is owner and developer of Aspen Grove Gardens, a six-acre botanical garden featuring 3,000 varieties of plants. A popular and well respected speaker and teacher, Robert has published articles in *Mother Earth News*, *Ontario Gardening* magazine, a monthly "Plant of the Month" column for the Ontario Rock Garden Society website, and local newspapers. He is also the author of the widely read blog GardenMyths.com which explodes common gardening myths; and GardenFundamentals.com, which provides gardening and garden design information. Robert also has a gardening YouTube channel called Garden Fundamentals.

Connect with Robert Pavlis

You can connect with me through social media by leaving comments in one of the following:
- gardenmyths.com
- gardenfundamentals.com
- youtube.com/Gardenfundamentals1

The best way to reach me directly is through my Facebook Group: https://www.facebook.com/groups/GardenFundamentals, where I answer questions on a daily basis.

Also by the Author

Building Natural Ponds

Building Natural Ponds is the first step-by-step guide to designing and building natural ponds that use no pumps, filters, chemicals or electricity and mimic native ponds in both aesthetics and functionality. Highly illustrated with how-to drawings and photographs.

For more information and ordering details, visit: BuildingNaturalPonds.com

Garden Myths

If you enjoyed this book you may also like my other books, *Garden Myths Book 1* and *Garden Myths Book 2*. Each one examines over 120 horticultural urban legends.

"Turning wisdom on its head, Robert Pavlis dives deep into traditional garden advice and debunks the myths and misconceptions that abound. He asks critical questions and uses science-based information to understand plants and their environment. Armed with the truth, Robert then turns this knowledge into easy-to-follow advice." Details about the books can be found atgardenmyths.com/garden-myths-book-1. They are available from Amazon and other online outlets.

Soil Science for Gardeners

"Robert Pavlis, a gardener for over four decades, debunks common soil myths, explores the rhizosphere and provides a personalized soil fertility improvement program in this three-part popular science guidebook." Coverage includes:

- Soil biology and chemistry and how plants and soil interact
- Common soil health problems, including analyzing soil's fertility and plant nutrients
- The creation of a personalized plan for improving your soil fertility, including setting priorities and goals in a cost-effective, realistic time frame.
- Creating the optimal conditions for nature to do the heavy lifting of building soil fertility

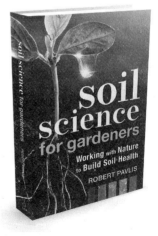

Written for the home gardener, market gardener and micro-farmer, *Soil Science for Gardeners* is packed with information to help you grow thriving plants.

ABOUT NEW SOCIETY PUBLISHERS

New Society Publishers is an activist, solutions-oriented publisher focused on publishing books to build a more just and sustainable future. Our books offer tips, tools, and insights from leading experts in a wide range of areas.

We're proud to hold to the highest environmental and social standards of any publisher in North America. When you buy New Society books, you are part of the solution!

At New Society Publishers, we care deeply about *what* we publish—but also about *how* we do business.

- All our books are printed on 100% **post-consumer recycled paper**, processed chlorine-free, with low-VOC vegetable-based inks (since 2002). We print all our books in North America (never overseas)

- Our corporate structure is an innovative employee shareholder agreement, so we're one-third employee-owned (since 2015)

- We've created a Statement of Ethics (2021). The intent of this Statement is to act as a framework to guide our actions and facilitate feedback for continuous improvement of our work

- We're carbon-neutral (since 2006)

- We're certified as a B Corporation (since 2016)

- We're Signatories to the UN's Sustainable Development Goals (SDG) Publishers Compact (2020–2030, the Decade of Action)

To download our full catalog, sign up for our quarterly newsletter, and to learn more about New Society Publishers, please visit newsociety.com

ENVIRONMENTAL BENEFITS STATEMENT

New Society Publishers saved the following resources by printing the pages of this book on chlorine free paper made with 100% post-consumer waste.

TREES	WATER	ENERGY	SOLID WASTE	GREENHOUSE GASES
46 FULLY GROWN	3,700 GALLONS	19 MILLION BTUs	150 POUNDS	19,700 POUNDS

Environmental impact estimates were made using the Environmental Paper Network Paper Calculator 4.0. For more information visit www.papercalculator.org